The Diplomacy of Food Dependence

Felice Farina

The Diplomacy of Food Dependence

Japan–United States Relations, 1945–1973

PETER LANG
New York · Berlin · Bruxelles · Chennai · Lausanne · Oxford

Bibliographic information published by the Deutsche Nationalbibliothek. The German National Library lists this publication in the German National Bibliography; detailed bibliographic data is available on the Internet at http://dnb.d-nb.de.

Library of Congress Cataloging-in-Publication Data
Names: Farina, Felice, 1986- author.
Title: The diplomacy of food dependence : Japan-United States relations, 1945-1973 / Felice Farina.
Description: First edition. | New York, NY : Peter Lang Publishing, Inc., 2025. | Includes bibliographical references and index.
Identifiers: LCCN 2025015841 (print) | LCCN 2025015842 (ebook) | ISBN 9783034358057 (hardback) | ISBN 9783034358033 (ebook) | ISBN 9783034358040 (epub)
Subjects: LCSH: Food industry and trade--Japan--History--20th century. | Food habits--Japan--History--20th century. | Agriculture and state--Japan--History--20th century. | Nutrition policy--Japan--History--20th century. | Japan--Foreign economic relations--United States x History--20th century. | United States--Foreign economic relations--Japan x History--20th century.
Classification: LCC HD9016.J42 F37 2025 (print) | LCC HD9016.J42 (ebook)
LC record available at https://lccn.loc.gov/2025015841
LC ebook record available at https://lccn.loc.gov/2025015842

Cover design by Peter Lang Group AG

ISBN 978-3-0343-5805-7 (print)
ISBN 978-3-0343-5803-3 (ePDF)
ISBN 978-3-0343-5804-0 (ePub)
DOI 10.3726/b22828

Published by Peter Lang Publishing Inc., New York (USA)
info@peterlang.com - www.peterlang.com

This publication has been peer reviewed.

Table of Contents

Acknowledgments

This book is the result of years of research conducted at the Department of Asian, African, and Mediterranean Studies of the University of Naples L'Orientale, where I also completed my Ph.D. I am deeply grateful to my department for providing a stimulating academic environment that has been essential for my growth as a scholar. I am profoundly grateful to my Ph.D. advisor, Professor Noemi Lanna, whose guidance has been far more than just academic. Her unwavering support, intellectual generosity, and keen insights have shaped not only this work but also my growth as a scholar. She has read this manuscript with patience and dedication, always offering thoughtful advice and encouragement. Her belief in my research, even in moments of doubt, has been a source of strength and motivation. I cannot thank her enough for the invaluable role she has played in this journey. I am also profoundly thankful to Professor Nissim Otmazgin for offering me the opportunity to be a visiting researcher at the Louis Frieberg Center for East Asian Studies at the Hebrew University of Jerusalem. The resources and scholarly community I encountered there greatly enriched my research and provided access to critical material. A special acknowledgment goes to Professor Ishida Masayoshi, who welcomed me as a visiting researcher at the College of Gastronomy Management at Ritsumeikan University in Minami Kusatsu, Japan. The year I spent there was crucial in gathering essential sources and refining my study. Finally, I would like to thank the anonymous reviewers for their valuable insights and constructive suggestions, as well as the editors and staff at Peter Lang for their professionalism and support throughout the publication process.

List of Abbreviations

A.A. System	Automatic Approval System
A.C.V.A.	American Council of Voluntary Agency
C.I.A.	Central Intelligence Agency
C.L.O.	Central Liaison Office
C.P.A.C.	Centro de Pesquisa Agropecuária dos Cerrados
C.R.A.L.O.G.	Committee of Relief Agencies Licensed to Operate in Germany
E.R.O.A.	Economic Rehabilitation in Occupied Areas
F.A. System	Fund Allocation System
F.A.O.	Food and Agriculture Organization
F.E.C.	Far Eastern Commission
F.R.U.S.	Foreign Relations of the United States
G.A.R.I.O.A.	Government Aid and Relief in Occupied Areas
G.H.Q.	General Headquarters
I.E.F.C.	International Emergency Food Council
JA.C.A.R.	Japan Center for Asian Historical Records
J.I.C.A.	Japan International Cooperation Agency
L.A.R.A.	Licensed Agencies for Relief in Asia
M.A.F.F.	Ministry of Agriculture, Forestry, and Fisheries
M.I.T.I.	Ministry of International Trade and Industry
M.O.F.A.	Ministry of Foreign Affairs
M.S.A.	Mutual Security Act
N.H.K.	Nippon Hōsō Kyōkai (Japan Broadcasting Corporation)
N.R.I.	Nomura Research Institute
N.R.S.	Natural Resources Section
O.D.A.	Official Development Assistance
O.E.C.D.	Organization for Economic Cooperation and Development

P.H.W.	Public Health and Welfare Section
P.L.480	Public Law 480
PRO.DE.CER.	Projeto de Desenvolvimento do Cerrado
S.C.A.P.	Supreme Commander of the Allied Powers
S.W.N.C.C.	State-War-Navy Coordinating Committee
U.N.R.R.A.	United Nations Relief and Rehabilitation Agency
U.S.D.A.	United States Department of Agriculture
U.S.S.B.S.	United States Strategic Bombing Survey

Disclaimer

All translations in this book are my own unless otherwise noted.

Japanese names in this book follow the Japanese convention in which the surname comes before the given name.

For transliterations from Japanese, I use the modified Hepburn romanization system.

Introduction

A historical view of Japan's food dependence

The stability of food supplies has been a consistent concern in Japanese diplomacy since the country opened its ports to world trade in the second half of the nineteenth century when Japan signed the Ansei Five-Power Treaties, marking the beginning of its modern international history.[1] The treaties, signed with the United States, the United Kingdom, France, Russia, and the Netherlands, contained, among other provisions, a ban on the export of major cereals, notably rice and wheat, while maintaining the obligation to provide food to the crews and passengers of foreign ships and to all other foreigners present in Japan. This kind of protectionist practice, whose aim was to avert food shortages, was widespread across East Asia. However, in the mid-1870s, Japan embarked on a significant policy shift by lifting its export restrictions, thereby initiating a phase of expanded exports. This expansion was facilitated by the increased rice production, made possible through the expansion of cultivated land, the adoption of fertilizers, and the implementation of new irrigation systems.[2] One thing that must be noted is that, particularly in rural areas, rice did not occupy the position of the main staple food grain as it does today, but it was commonly consumed mixed with wheat and other cereals. Due to the relatively low proportion of rice in the staple diet, the growth in rice production did not immediately correspond to a proportional increase in its consumption. This led to a surplus of rice, which

---

1 Michael R. Auslin, *Negotiating Imperialism. The Unequal Treaties and the Culture of Japanese Diplomacy*, Cambridge, Massachusetts, Harvard University Press, p. 1.

2 Penelope Francks, *Rural Economic Development in Japan. From the Nineteenth Century to the Pacific War*, London, Routledge, 2006.

was then used for export.[3] However, in the 1890s, the combination of a significant increase in population and a rise in rice consumption as a staple food led to a gradual shift toward a shortage in the supply-demand balance. As a result, meeting domestic demand through local production alone became increasingly difficult. This shortfall had to be compensated for through trade, generating a strong dependence on foreign supplies, a situation that would characterize Japan for years to come and that will be referred to as the "food problem" (*shokuryō mondai*). According to Japanese historian Ōmameuda Minoru, it is possible to identify four main phases in Japan's international trade of rice and other key agricultural foods during the period from 1890 to 1945. Between 1890 and 1920, the advent of industrialization and urbanization, together with a considerable increase in population, led to a notable increase in rice consumption. As a result, domestic rice production proved insufficient to meet national demand, necessitating food imports (*shokuryō yunyū*) from Southeast Asia or "food inflows" (*shokuryō inyū*) from Japan's colonized territories, that is, Taiwan and Korea. The disruption of foreign rice (*gaimai*) import system following the end of World War I resulted in a surge in rice prices, exacerbating labor discontent and social discord, ultimately leading to the outbreak of the "Rice Riot" (*Kome sōdō*) in 1918. Governmental efforts during the 1920s aimed to address this crisis by bolstering domestic rice production, both within Japan itself and in its colonial territories. This initiative sought to achieve full self-sufficiency within the Japanese empire and reduce dependence on *gaimai* imports. The third phase, on the other hand, marks the period following the seismic economic downturn triggered by the Great Depression of 1929. The international economic crisis had profound and far-reaching repercussions on the Japanese agricultural sector, destabilizing rural livelihoods and exacerbating existing vulnerabilities. The response to the agricultural depression quickly evolved into a pressing political issue, particularly in the wake of the collapse of civilian governments in late 1931 and the rise of militarism. While the government continued to emphasize the goals of food self-sufficiency and price stabilization, its primary focus began to shift toward the implementation of comprehensive measures for "rural relief" (*nōson kyūsai*), designed to alleviate the widespread hardship afflicting the countryside. The fourth and final phase pertains to the period of the war between the outbreak of the Second Sino-Japanese War in 1937 and the end of the Pacific War in 1945. During this time, all efforts were focused on achieving total state control over the economy.

---

3 Ōmameuda Minoru, *Kindai nihon no shokuryō seisaku. Taigai izon beikoku kyōkyū kōzō no hen'yō* [Food policies in modern Japan: Transformation of the structure reliant on imported rice supply], Kyoto, Minerva Shobō, 1993, p. 11.

To this end, the 1939 Rice Distribution Control Law (*Beikoku haikyū tōsei hō*) and the 1942 Food Control Law (*Shokuryō kanri hō*) were enacted, granting the state complete authority over food production and distribution. Additionally, a program was initiated for agricultural development in Manchuria, involving the emigration of Japanese farmers. Despite the program's incomplete implementation, food supplies from Manchuria swiftly became one of the primary sources of sustenance for the Japanese populace. However, as the war intensified and the Japanese military faced defeats and territorial losses, the procurement of supplies from the empire became increasingly difficult, leading to widespread hunger among the Japanese people.[4]

The defeat in 1945 put a definitive end to the aspirations for a self-sufficient empire, leaving Japan to solely rely on domestic production to sustain its population. However, confronted with the inability to secure a minimal food supply essential for subsistence and the consequential risks to the nation's stability, the United States extended assistance to Japan by providing the necessary food through food aid. With the arrival of food aid, the two nations developed an increasingly interdependent relationship, which continued long after the Occupation ended in 1952, with Japan serving as the primary outlet for American agricultural surpluses and the United States serving as the major market for Japan's resurgent industry. In a brief span of time, the United States would provide approximately 90 percent of all agricultural imports to Japan.

The Japanese food dependence on the United States has been the subject of several studies. In the case of Japanese-language literature, the prevailing narrative portrays Japan as having been trapped by the so-called "American wheat strategy" (*Amerika komugi senryaku*), whose aim was to support domestic agricultural production and allocate its surpluses overseas through food aid and export. Japan, as a defeated country occupied by the Americans, thus found itself forced to accept food aid first and then to import agricultural products from the United States. This narrative gained prominence through N.H.K.'s March 1978 special program "The Shadow of the Stars and Stripes Flag on the Dining Table: Postwar History of Rice and Wheat" (*Shokutaku no kage no seijōki: kome to mugi no sengoshi*), authored by Takashima Teruyuki, who also published the book "The Invasion of Japan: The American Wheat Strategy" (*Nihon shinkō: Amerika komugi senryaku*) in 1979. According to Takashima, the American strategy can be understood as a response to lobbying efforts, particularly from wheat producers in Oregon, aimed at capitalizing on the sale of surplus agricultural goods to Japan. The result of this strategy was the inundation of the Japanese market with cheap

[4] Ibidem.

agricultural products, resulting in detrimental effects on domestic production. Central to this strategy were programs aimed at influencing shifts in Japanese dietary preferences, exemplified by initiatives like school meals and kitchen cars. These endeavors sought to promote greater consumption of American products while steering away from traditional staples like rice. Over the years, other scholarly works have emerged that corroborate the idea of Japan as a victim of the "American wheat strategy," exerting a profound influence not only on the Japanese academic discourse, but also on mass media and public opinion.[5] The notion that Japan's dependence on imports from the United States stems from American food strategy has found resonance in Western literature as well. Compared to Japanese-language literature, it is possible to observe a greater emphasis on the international context and the strategic objectives of the United States and its allies from the perspective of the Cold War. David Moen argues that Japan's dependence from the United States was driven by geopolitical and economic factors, particularly U.S. strategic interests and Japan's alignment with American foreign policy during the Cold War. He claims that the United States used food as a political weapon for its strategic and economic interests and played a crucial role "in helping to engineer the postwar agricultural debacle in Japan."[6] Moen also emphasizes the role of the Japanese government in reinforcing the country's dependence on U.S. agricultural imports. While he never uses the expression "wheat strategy," Moen portrays Japan's agricultural policy as a deliberate choice that aligned with U.S. geopolitical and economic interests, rather than an unavoidable consequence of market forces. Building on Takashima's analysis, historian Nathan Hopson asserts that the American food aid, the kitchen cars program, and the school lunch program promoted in the 1950s, must be understood within the context of the Cold War. He argues that the primary objectives of American food aid were "promoting American agricultural security and stability, promote international trade in U.S. agricultural commodities, and bolster the economic development of Cold War American allies and friendly nations." These programs, he adds, were

5 See: Ishikawa, Hirotomo, *Kokumotsu mejā: Shokuryō senryaku no "kage no shihaisha"* [Grain Majors: Rulers of Food Strategy], Tokyo, Iwanami Shoten, 1981; Kishi, Yasuhiko, *Shoku to nō no sengoshi* [The Postwar History of Food and Agriculture], Tokyo, Nihon keizai shinbunsha, 1996; Kawai, Kazushige, *Osorubeki "yunyū kome" senryaku: WTO kyōtei kara kome to tanbo wo mamoru tame ni* [The Terrifying "Imported Rice" Strategy: Protecting Rice and Rice Fields from WTO agreement], Tokyo, Gōdō, 2000; Suzuki, Takeo, *Amerika no komugi senryaku to nihonjin no shokuseikatsu* [The American Wheat Strategy and Japanese Food Habits], Tokyo, Fujiwara Shoten, 2003; Suzuki, Nobuhiro, *Shoku no sensō. Beikoku no wana ni ochita Nihon* [The War of Food: Japan Caught in the American Trap], Tokyo, Bunshun, 2013.

6 Darrell G. Moen, "The Postwar Japanese Agricultural Debacle," in *Hitotsubashi Journal of Social Studies*, vol. 31, n. 1, pp. 29–52, 1999.

instrumental in teaching Japan to accept and consume American products.[7] In a similar vein, George Solt, in his comprehensive and compelling exploration of the history of ramen, claims that wheat was "an important tool in America's Cold War strategy" and that there were "clear political pressures exerted upon Japanese government officials by American representatives of agribusiness to absorb immense amounts of U.S. wheat."[8] Not only historians but also scholars of international food regimes have explored Japan's dependence through a structuralist lens. They argue that this dependence is a result of the global restructuring of food production in the postwar era, which led to the establishment of a food regime centered around American agricultural surpluses.[9]

Recent contributions from Japanese scholars, particularly exemplified by the work of Professor Itō Atsushi from Kyoto University, have begun to challenge this paradigm. More specifically, Itō has highlighted a significant gap between the intentions and results of the "American Wheat Strategy." During the active promotion of the wheat strategy, per capita wheat consumption decreased while rice consumption increased, and wheat imports from the United States were declining. Furthermore, the United States primarily exported semi-hard and soft wheat varieties, which were hardly used for bread but rather for noodles such as ramen and udon, while Canada was the main source for hard wheat imports. An aspect that has also been highlighted by the historian Barak Kushner.[10] Moreover, besides wheat, Japan also imported other agricultural products from the United States, indicating that the strategy was not solely focused on wheat exports. One of the main goals of the United States was to boost also exports of rice and soybeans, both essential ingredients in the traditional Japanese diet. Soybean exports gained importance whereas the plan to export rice ultimately failed, due to competition from rice imported from Southeast Asia. However, according to Itō, the very fact that the United States aimed to develop a market for these products in Japan challenges one of the key assumptions of the "wheat strategy," which suggested that the U.S. intended to push for a shift in the Japanese diet. Instead,

---

7 Nathan Hopson, "Ingrained Habits: The 'Kitchen Cars' and the Transformation of Postwar Japanese Diet and Identity," in *Food, Culture & Society*, vol. 23 n. 5, pp. 589–607.

8 George Solt, *The Untold History of Ramen: How Political Crisis in Japan Spawned a Global Food Craze*, Berkley, University of California Press, 2014.

9 Harriet Friedmann, "The Political Economy of Food: A Global Crisis," in *New Left Review*, vol. I n. 197: 29–57, 1993; Harriet Friedmann and Philip McMichael, "Agriculture and the State System: The Rise and Decline of National Agricultures, 1870 to the Present," in *Sociologia Ruralis* vol. 29 n. 2, pp. 93–117, 1989; Philip McMichael, "A Food Regime Genealogy," in *The Journal of Peasant Studies*, vol. 36 n. 1, pp. 139–69, 2009.

10 Barak Kushner, *Slurp! A Social and Culinary History of Ramen—Japan's Favorite Noodle Soup*, Leiden, Global Oriental, 2012, p. 211.

the significant increase in wheat imports occurred after the 1961 Agricultural Basic Law, which aimed to ensure food self-sufficiency domestically. He also refutes the idea the United States intended to develop a Japanese market for U.S. wheat through the school lunch program.[11]

The present monograph aligns itself with this evolving scholarly discourse. While accepting the idea that the Cold War international context was crucial in the development of this dependent relationship, the main aim is to challenge the conventional narrative, particularly evident in Japanese scholarship, that portrays Japan as a passive victim of a deliberate American strategy designed to enforce dependence in the aftermath of the Pacific War. By adopting a more nuanced and critical perspective, this work aspires to dismantle the simplistic dichotomy of domination and submission in the bilateral relationship between the two countries and instead shed light on the intricate dynamics underpinning Japan's postwar reliance on food imports from the United States. To this end, this work will focus mainly on the diplomatic dimension of the food issue. Through the analysis of diplomatic archives and contemporaneous newspapers from both Japanese and American sources, this study aims to demonstrate that Japan's reliance on food imports from the United States did not arise solely as the coerced consequence of defeat. Rather, it was the product of a sophisticated interplay of negotiations, conflicts, and strategic calculations conducted by both nations, with each seeking to secure and promote its own national interests. We will also attempt to highlight how not only the general national and international interests but also the individual perspectives of the key figures involved in the negotiation processes played a crucial role. In this way, this research contends

[11] Itō, Atsushi, "Dai ichi ji PL480 taitoru I kyōtei o meguru nichibei kōshō, 1954–55" [Negotiations between Japan and the United States regarding the First PL480 Title I Agreement, 1954–55], in *Rekishi gaku kenkyū*, vol. 3, pp. 1–15, 2021; "Dai ni ji PL480 taitoru I kyōtei o meguru nichibei kōshō, 1955–56" [Negotiations between Japan and the United States regarding the Second PL480 Title I Agreement, 1955–56], in *Shakai keizai-shi gaku*, vol. 88 n. 1, pp. 33–55, 2022; "Dai san yon ji PL480 taitoru I kyōtei o meguru nichibei kōshō, 1956–57" [Negotiations between Japan and the United States regarding the Third and Fourth PL480 Title I Agreement, 1956–57], in *Shakai keizai-shi gaku*, vol. 89 n. 1, pp. 27–52, 2023; "PL480 taitoru II o meguro nichibei kōshō—Gakudō-fuku keikaku no dannen to gakkō kyūshoku muke zōyo no seiritsu" [The Negotiations between the U.S. and Japan concerning PL 480 Title II: The Abandonment of a School Children's Clothing Program and the Formation of a Grant for a School Lunch Program], in *Nōgyō keizai kenkyū*, vol. 92 n. 2, pp. 165–77, 2020; "M.S.A. yojō nōsanbutsu kyōtei o meguru nichibei kōshō 1953–60—Dai 402 jō kyōtei kōshō ni chakumoku shite" [The Negotiations between the U.S. and Japan Concerning Surplus Agricultural Commodities Agreements under the Mutual Security Act, 1953–60: A Focus on Section 402 Negotiations], in *Nōgyō keizai kenkyū*, vol. 94 n. 3, pp. 208–13, 2022.

that the relationship of interdependence between Japan and the United States was not unilaterally imposed but was carefully constructed through mutual accommodations that addressed a range of economic, social, political, and security imperatives, as well as personal perspectives. By revealing the agency and active participation of Japanese policymakers in shaping the country's economic and agricultural trajectories, this monograph repositions Japan as a key actor rather than a passive recipient in this historical process. This study aims to demonstrate how Japan, through the key figures involved in decision-making and negotiations on food-related matters, consistently adopted a proactive stance in safeguarding the nation's economic, political, social, and security interests. This approach mirrored that of its counterparts, often leading to significant conflicts and prolonged negotiations; however, in the end, it proved mutually advantageous for both countries: it facilitated the export of American agricultural surpluses to Japan while opening American markets to Japanese industrial goods, thus fostering economic growth on both sides. At the same time, it addressed critical social challenges, such as alleviating food scarcity in Japan, and served broader political and security objectives, including the containment of Communism and the strategic rearming of Japan within the framework of Cold War geopolitics.

Ultimately, this work aspires to offer a richer, more comprehensive understanding of the complexities surrounding postwar Japan-United States relations, emphasizing the multifaceted interplay of power, pragmatism, and mutual benefit that defined this pivotal era in global history.

Structure and sources of the book

The analysis presented in this monograph spans the period from 1945 to 1973, a transformative epoch that charts the evolution of Japan's dependence on food imports from the United States. This timeline starts with the implementation of Operation Starvation during the Pacific War's final stages and concludes with the soybean embargo in the summer of 1973. Within this nearly three-decade scope, three pivotal historical moments are identified, each reflecting critical shifts in the U.S.-Japan food supply relationship. These moments structure the three chapters of the study.

The first chapter examines the immediate postwar period, beginning with the United States' implementation of Operation Starvation in spring 1945, extending through Japan's early Occupation under Allied control, and concluding in summer 1946. This phase is characterized by a punitive approach imposed by the Allies, particularly the United States, which aimed to limit Japan's access to international food markets, thereby forcing reliance on domestic agricultural production. The analysis highlights Japan's urgent lobbying to regain access to international trade, driven by severe domestic food shortages and fears of widespread social unrest. Simultaneously, debates within the General Headquarters (G.H.Q.) revealed stark divergences among the Allies: while the United Kingdom and Commonwealth nations favored stringent restrictions, U.S. leaders, including General Douglas MacArthur and George C. Atcheson Jr., advocated for leniency. These American actors ultimately persuaded Washington to provide food aid, emphasizing its strategic importance in stabilizing Japan and preempting unrest that could undermine the Occupation's goals.

The second chapter investigates the arrival of American food assistance programs starting in summer 1946, marking the resolution of Japan's immediate food crises. Programs such as L.A.R.A., G.A.R.I.O.A., E.R.O.A., and, later, M.S.A. and P.L.480 shaped Japan's recovery during both the Occupation and the post-independence period following the 1951 San Francisco Peace Treaty. The chapter identifies three key imperatives driving these programs: addressing Japan's humanitarian food crisis, managing American agricultural surpluses, and aligning Japan with Western political and economic interests to counter the spread of Communism. Contrary to interpretations framing these efforts as coercive, the research argues that Japanese authorities actively sought food aid to rapidly alleviate shortages, foster social stability, and sustain industrial growth through enhanced trade relations with the United States. This pragmatic collaboration between the two nations laid the groundwork for deeper economic ties.

The final chapter explores the evolving tensions in United States-Japan food relations, beginning with debates over the liberalization of Japan's agricultural market in the late 1950s. Pressured by both American and European partners to conform to global trade norms, Japan faced also internal conflicts. The Ministry of Agriculture, Forestry and Fisheries (M.A.F.F.) resisted liberalization, aiming to safeguard domestic production, while the Ministry of International Trade and Industry (M.I.T.I.) advocated for greater openness to expand access to international markets for Japanese exports. By the 1970s, these tensions reached a turning point with the Nixon administration's imposition of the so-called "soybean embargo," a unilateral restriction on agrifood exports, enacted in order to mitigate an inflationary crisis. While the embargo had limited immediate effects on Japan's food supply, it significantly influenced the country's perception of its dependence on imports, reviving fears of food insecurity. In response, Japan undertook measures to reduce its dependence on American agricultural imports, focusing on diversifying its food sources through investments in developing nations. This marked a pivotal shift in Japan's approach to ensuring long-term food security.

This monograph's narrative is constructed using a wealth of archival materials. Primary sources include the *Records of the U.S. Department of State Relating to the Internal Affairs of Japan, 1950–1954* (microform), the *Japan National Diet Library Digital Collections*, the *National Archives of Japan—Digital Archive*, the *Foreign Relations of the United States* series, and materials from the U.S. National Archives and Records Administration. Supplementary insights are derived from contemporaneous newspapers, including *The New York Times*, *Asahi Shinbun*, *Japan Times*, and *Yomiuri Shinbun*. These sources collectively shed light on the multifaceted interplay of economic, political, and social forces that shaped the United States-Japan food relationship during this critical period. By weaving these threads together, the monograph seeks to contribute to a more nuanced understanding of the historical interdependence between the United States and Japan, emphasizing mutual agency and the complex calculus underlying their food supply dynamics.

CHAPTER 1

Punishing Japan: The food problem in 1945–46

1.1 Introduction

August 15, 1945, witnessed the conclusion of one of the most destructive conflicts in human history, as Japan acknowledged its defeat and acquiesced to the conditions of unconditional surrender that were enunciated in the Potsdam Declaration. This denouement marked the termination of a protracted war that had endured for 14 years, commencing with the Manchurian Incident of September 1931, and culminating in the deployment of atomic weapons on Hiroshima and Nagasaki in August 1945. For the first time, Japan was compelled to acknowledge a foreign conqueror and endure occupation by the Allied Forces, marking a significant juncture in the nation's history.

Upon their arrival in Japan in September 1945, the occupying forces were confronted with a nation in a state of disarray, with a landscape devastated by war and a population that was on the brink of starvation. The most pressing concern for the majority of Japanese was to procure sustenance for their basic needs, as the nation had been unable to maintain its agricultural and industrial production due to the effects of the war. As such, food scarcity became a major issue, and the task of addressing the basic needs of the population became one of the primary objectives of the occupying forces. But—as John Dower has brilliantly described in his Pulitzer-prize book *Embracing Defeat*[12]—hunger was not simply the consequence of the defeat. A majority of Japanese were already malnourished before the war ended. The inability to maintain the flow of foods and fertilizers from foreign and colonial sources, the necessity of ensuring adequate nutrition to the soldiers, the mobilization and conscription that reduced the available manpower

---

12 John W. Dower, *Embracing Defeat: Japan in the Wake of World War II*, New York: WW Norton & Company, 1999, p. 89–97.

in agriculture and other economic sectors were among the causes that pushed the government to limit the consumption of food, through rationing policies, leading to a deterioration in the quantity and quality of the diet of the Japanese people. The reduction in food consumption is well illustrated by caloric consumption. If during the year 1941 the average per capita consumption of food approximated 2,000 calories per day, by 1945 it declined to only 1,680 calories.[13]

In this chapter, we will focus on the last months of the Pacific War and on the first year of the Allied Occupation of Japan, which we can consider as the worst period in modern and contemporary history of the country from the point of view of nutrition. In particular, as regards the last months of the war, we will start by investigating the deleterious effects of the war events on the production and distribution of food in Japan and analyzing the American strategy of blockade, named "Operation Starvation," whose aim was to prevent any food supply from outside to the archipelago. We will then focus on the first year of the Occupation and examine how, even during this period, despite the war having ended and Japan being defeated, the Allied Forces maintained the notion that the entire Japan, including its civilian population, should be punished for the crimes committed and that no form of food aid should be extended to them. The Occupation Forces had been mandated not to take any action in favor of the economic rehabilitation of Japan and to pursue only the goals of demilitarization and democratization of the country. However, we will also see that, in light of the severe food shortage, General Douglas MacArthur and his advisor George C. Atcheson Jr. quickly realized that without addressing the food problem, the objectives of the Occupation would be jeopardized, rendering American efforts futile. The issue of resolving Japan's food problem thus saw a strong contraposition between, on one side, MacArthur and Atcheson, who were in favor of accommodating the Japanese government's requests for food aid, and on the other side, the remaining members of the Occupation Forces, particularly the United Kingdom and the representatives of the Commonwealth countries, who continued to support a punitive stance against any form of food assistance. We will show how the United States ultimately prevailed over the other Allies. The restoration of food imports, the resumption of open-sea fishing, and the arrival of the first food aid marked the end of the punitive stance toward Japan, a stance that had started with Operation Starvation.

---

13 The United States Strategic Bombing Survey (hereafter cited as U.S.S.B.S..), *The Japanese Wartime Standard of Living and Utilization of Manpower*, Washington D.C.: Manpower, Food and Civilian Supplies Division, 1947, p. 100.

1.2 The effects of the war on Japan's food production

Before examining the American strategy against Japanese food supplies, it is essential to understand the food situation in Japan during the war and how the country reached such a critical state of food scarcity. The impact of the war was detrimental to all aspects of Japan's food situation, and it was impossible to maintain domestic food production on the prewar level. In particular, the decline in Japanese agricultural production between 1941 and 1945 was considerably influenced by the scarcity of physically capable young male workers, alongside a reduction in the availability of fertilizers, and other conditions induced by the war conditions that served to diminish agricultural yields.[14] Moreover, during the 1930s, Japan aggressively pursued self-sufficiency within its empire, increasingly relying on agricultural products from its colonies, such as Taiwan and Korea, as well as other controlled territories like Manchuria. These regions effectively became Japan's "agricultural appendages," integral to sustaining the nation's food supply.[15] As the conflict intensified, it became increasingly difficult to transport food from these territories, exacerbating an already dire situation for the Japanese population.

The shortage of agricultural labor during the war can be attributed primarily to the pressing demand for workers in the war industry. It has been estimated that nearly two million rural workers abandoned their farms to work in war plants.[16] But it was military conscription to deprive agricultural sector of a significant portion of its most productive workers. According to the Japanese War Ministry, approximately 23 percent of military recruits during the Pacific War were drawn from agriculture.[17] Recognizing the serious consequences of labor scarcity in agriculture, the Japanese government implemented several measures between 1941 and 1944 to alleviate the problem. These measures included enforcing restrictions to keep farmers in their occupations, granting deferments for agricultural workers from mobilization into war industries, promoting the use of draft animals and cooperative farming initiatives, mobilizing students for agricultural work, and organizing special mobile works corps.[18] Despite these efforts, the unfavorable course of the war hindered their effectiveness. By January

[14] U.S.S.B.S., *The Japanese Wartime Standard of Living*, p. 9.

[15] Samuel Pao-San Ho, "Colonialism and Development: Korea, Taiwan, and Kwantung," in *The Japanese Colonial Empire*, 1895–1945, ed. Raymond H. Myers and Mark R. Peattie, Princeton: Princeton University Press, 1984, p. 350.

[16] Ibid., p. 10.

[17] Bruce F. Johnston, *Japan Food Management in World War II*, Wisconsin: Stanford University Press, 1953, p. 94.

[18] U.S.S.B.S., *The Japanese Wartime Standard of Living*, p. 10.

1945, it became necessary to increase the conscription of agricultural workers in preparation for the expected invasion of the home islands. As a result, approximately 1.6 million men had been drawn from Japanese agriculture to serve in the armed forces by the end of the war.[19]

Another important factor that had a negative impact on Japan's agricultural productivity was the decline in the production of fertilizers. This decline can be primarily attributed to the difficulty in importing the essential chemical materials required to produce fertilizers, specifically phosphorite and potassium salts.[20] Prior to the war, Japan had the capacity to produce approximately 3,500,000 tons of fertilizers annually. However, this output experienced a substantial decline during the war years. By 1943, annual fertilizer production had plummeted to 1,665,000 tons, and by the war's end, it had further decreased to a mere 630,000 tons.[21]

Furthermore, the war significantly hindered Japan's ability to import large quantities of agricultural and food products from overseas to compensate domestic shortages. By 1943, the American forces had sunk not only 22 Japanese warships but also 296 merchant vessels, severing Japan's primary sources of rice imports from Southeast Asia.[22] Moreover, the number of military personnel stationed in Japan surged from one million to 3.5 million between 1941 and 1945. Consequently, the amount of rice that these forces needed increased from 161,000 to 744,000 tons, surpassing the entire quantity of rice smuggled past the American blockade in 1943.[23]

The decimation of the Japanese merchant fleet also compromised the domestic distribution of food from rural areas to urban populations. The majority of domestic food production was concentrated in Hokkaidō and Kyūshū, necessitating shipping to transport agricultural products to the major urban centers in Honshū. However, even when food was intended for consumption within the same island where it was produced, the mountainous terrain of Japanese archipelago made it more convenient to transport goods by rail to the nearest port and then ship them along the coast to the cities. Consequently, the urban population was left with a minimal portion of an increasingly diminished harvest.[24]

The impact of war mobilization and economic conversion extended beyond the agricultural sector, affecting the fishing industry as well, which has historically

---

19 Johnston, *Japan Food Management*, p. 94.
20 U.S.S.B.S., *The Japanese Wartime Standard of Living*, p. 10.
21 Ibidem.
22 Johnston, *Japan Food Management*, p. 169.
23 Ibid., p. 140 and p. 152.
24 Lizzie Collingham, *The Taste of War: World War Two and the Battle for Food*, London: Penguin U.K., 2011, pp. 286–87.

played a pivotal role in providing food for the Japanese population. Similar to agriculture, the initial consequence experienced by the fishing industry was a shortage of manpower due to the significant enlistment of men into military service. The number of individuals engaged in fishing peaked at 1.1 million in 1936 but steadily declined following the outbreak of the Pacific War in 1941, reaching only 721,000 by 1945.[25] Furthermore, the requisitioning of a substantial number of larger powered boats for military purposes directly and severely impacted the fishing sector.[26] The scarcity of fishing supplies, particularly fuel, cotton, and hemp required for the production of ropes and fishing nets, also contributed to a significant 50 percent decline in the fish catch, further exacerbating the nutritional value of the wartime diet.[27]

In this circumstance, it became evident that implementing some form of food rationing and regulating food prices was necessary. Following the outbreak of Sino-Japanese War in 1937, the Japanese government approved a number of measures to control food production, distribution and consumption, which were subsequently incorporated into the 1942 Food Control Law (*Shokuryō kanri-hō*). The resulting Food Control System (*Shokuryō kanri seido*) abolished the previous system of distributing staple foodstuffs through designated merchants and instead established quasi-governmental food control corporations in each prefecture to undertake this role. This system relied on three primary bureaucratic regulations: (1) controlling the mechanisms by which the price of rice was determined at all levels of the distribution system; (2) managing the distribution of rice from producers to consumers by regulating rice traders; and (3) exerting government control over external trade involving rice.[28] The level of public control over food distribution was so pervasive that it has been considered to have more in common to that of planned economies rather than of market-based ones.[29]

During the period from 1941 to 1945, the daily calorie intake of the Japanese population was reduced from approximately 2,000 calories per day, which was already below the level required for a satisfactory diet, to less than 1,700 calories, a dangerously low quantity approaching subsistence levels.[30] The inhabitants of major cities such as Tokyo, Yokohama, Nagoya, Osaka, and Kobe, were the hardest

25 Johnston, *Japan Food Management*, p. 130.
26 Ibid., pp. 129–30.
27 Ibid., pp. 130–1.
28 Penelope Francks, "Agriculture and the State in Industrial East Asia: The Rise and Fall of the Food Control System in Japan," *Japan Forum* 10, no. 1, January 1998, p. 3.
29 Ibid., p. 4.
30 U.S.S.B.S., *The Japanese Wartime Standard of Living*, p. 100.

hit because they were heavily reliant on the food rationing system established by the national government in 1940.[31]

1.3 Operation Starvation

Japan's difficulties to secure an adequate supply of food supply did not escape the attention of the United States. Recognizing this critical weakness, Fleet Admiral Chester Nimitz devised a mining campaign against Japanese shipping in July 1944, with the primary objective of cutting off external food supplies and pressuring the Japanese authorities into surrender. Initially, the proposal faced strong opposition from the Army, which advocated for a large-scale invasion of the Japanese islands. General Marshall, the Army's Chief of Staff, believed that a ground operation similar to Normandy landing would be more effective.[32] In contrast, Admiral Nimitz and General MacArthur, the senior commanders in the Pacific, disagreed with Marshall's perspective. They considered an invasion of Japan unnecessary, arguing that the destruction of its merchant fleet and the loss of access to oil, food, and other vital resources would sufficiently undermine Japan's offensive capabilities.[33] In an attempt to mediate between the two positions, the Committee of Operations Analysts recommended in September 1944 that Japan's defeat should primarily be achieved through strategic bombardment and blockade, followed by a potential invasion in late 1945.[34] On November 7, 1944, Admiral Nimitz addressed a personal memorandum to General Arnold and proposed a B-29 mining of Japanese home waters, suggesting a starting date between January and March 1945.[35] The B-29 mining campaign, named after the Boeing-designed heavy bomber used for high-altitude strategic bombing and mine deployment, started on March 27, 1945.

The mining operation had three main strategic objectives: (1) prevent the import of raw materials and food into Japan; (2) disrupt the supply and deployment of Japanese military forces; (3) disrupt internal marine transportation within

---

[31] Samuel Hideo Yamata, "The 'Food Problem' of Evacuated Children in Wartime Japan, 1944–1945," in *Food and War in Mid-Twentieth Century East Asia*, ed. Katarzyna J. Cwiertka, Farnham: Ashgate, 2013, pp. 131–2.

[32] Frederick M. Sallagar, *Lessons from an Aerial Mining Campaign (Operation "Starvation")*, Rand Report 1322-PR, April 1974, p. 16.

[33] Ibid., p. 17.

[34] Ibid., p. 21.

[35] Ibid., p. 27.

the Inland Sea.[36] The first point was considered crucial for diminishing the civilian will to support the war. In 1944, Japan relied on imports for approximately 20 percent of its total food consumption, which may be considered a small amount under normal circumstances. However, given the low nutrition standards at that time, a 20 percent reduction meant that a significant portion of the Japanese population would transition from a condition of subsistence to starvation. Hence, it is unsurprising that the operation was soon codenamed "Operation Starvation," despite the tendency to use less explicit names for military operations.[37]

The operation was conducted in five phases spanning from March 27 to August 5, 1945. During this period, more than 12,000 acoustic and magnetic mines were strategically deployed in the Western approaches to the Shimonoseki Strait and Inland Sea.[38] From the American perspective, the operation proved highly successful, resulting in the sinking or significant damage of 670 ships, including 65 warships. By the conclusion of the fifth phase, half of the remaining 1.5 million tons of the merchant marine fleet was damaged, leaving the remaining ships stranded in their harbors due to the mines laid down around Japan.[39] As a consequence, volumes of agricultural products from Manchuria were left rotting on the quays of the Korean ports, and any few ships that managed to evade the blockade did so in vain. Furthermore, the American mines disrupted coastal transport within Japan, exerting significant pressure on the inadequate rail system. Coupled with labor shortages at the docks, this resulted in food supplies rotting in the ship holds and on Japanese quaysides.[40] Even the soldiers who were fighting in the occupied territories, particularly those in the small Pacific islands, had to face shortages of food supplies. Reports suggest that soldiers stationed on the island of Nauru resorted to cruel and abhorrent acts of cannibalism against the local population when food became scarce following the American bombings.[41]

In addition to the bombings, another pivotal aspect of Operation Starvation was the implementation of a psychological warfare campaign. Early on in the mining operations, the Twentieth Air Force Planning conceived the idea of a psychological warfare campaign involving the dropping of propaganda leaflets that emphasized the impending specter of starvation resulting from the mining

---

36 U.S. Army Air Force, A-3 20th Air Force, "Starvation. Phase Analysis of Strategic Mining Blockade of the Japanese Empire," 1945, p. 3

37 Ibidem.

38 Ibid., pp. 21–33.

39 Collingham, *The Taste of War*, p. 211.

40 Ibidem.

41 Anonymous, "Cannibalism Laid to Enemy on Nauru," *The New York Times*, 15 September 1945.

blockade. The aim of this campaign was to instill a sense of helplessness among the Japanese population, thereby diminishing morale and eroding the enemy's will to continue fighting. Approval for the proposed psychological warfare campaign in conjunction with the mining operations was ultimately granted on May 29, 1945. The leaflets were carefully crafted to convey the following key messages:

- Japanese food supplies were rapidly being depleted by American fire raids.
- The Japanese homeland was unable to produce and distribute sufficient food to sustain its population.
- Importing an adequate amount of food was now impossible due to the minefields strategically placed by B-29 aircraft in Japanese home waters, blocking all harbors and shipping lanes.
- The Japanese government's insistence on risking ships through the minefields to supply iron and coal to the war effort resulted in substantial shipping losses.
- If this rate of loss had persisted, it would have been impossible to feed Japan's starving population even after achieving peace due to the scarcity of ships.
- Ultimately, if the current policy had continued, the entire Japanese population would have faced inevitable and disgraceful starvation.[42]

The Psychological Warfare Section of the Commander in Chief Pacific Fleet Headquarters prepared a set of four propaganda leaflets focusing on the subject of mine warfare. Four and one-half million of the first three leaflets were dropped widely over the entire Japanese homeland during the final phase of the mining campaign. The fourth leaflet was still in preparation when the war ended.[43] These leaflets included specific mentions of the B-29 Aerial Mining Campaign and its impact on Japan's food supplies. Here what it was written on the first of the three leaflets that were dropped on Japan:

> One of the excuses the militarists gave you for beginning this war was that it was essential to ensure the supply of food from overseas. And so you have been at war for many years. Now, however, the militarists are saying that your food supplies have sunk to an unprecedented low, and that, in the future, food rations will become smaller and smaller. The militarists are, however, concealing the fact that their blunders are sacrificing the people of Japan. What has happened to the ships which in peacetime transported provisions to Japan? Many ships carrying war materials have been sunk by

---

[42] U.S. Army Air Force, "Starvation," 1945, p. 48.
[43] Ibidem.

> American submarines and planes. Moreover, the B-29 have laid mines in the various harbors and coastal waters. A new threat to shipping has thus been added. How long will you allow the militarist want only to drive the people of Japan to starvation?[44]

The utilization of food scarcity and potential famine as a psychological warfare tactic by the Americans extended beyond the context of the B-29 mining campaign. The subject of reduced food provisions played a central role in the overarching American propaganda strategy employed in Japan throughout the war and was addressed in numerous additional leaflets disseminated across the archipelago. Among the notable examples was a widely recognized leaflet featuring an image depicting a substantial plate of sushi, accompanied by the following statement:

> Your island has been isolated and cut off from all aid and supplies. You have almost no food and are slowly starving to death. You are as human as we are and the thought of your hunger is far from pleasant. If you are hungry and wish to have good food, indicate that fact by displaying a large visible cross along the southeast intersection of the airfield runway. We will then be able to help you.[45]

It becomes evident that the aim of the psychological campaign was not solely to induce surrender but also to instill a sense of distrust toward the Japanese government and foster a positive perception of the American invaders.

At 8:16 a.m., on August 6, 1945, the *Enola Gay* dropped the first atomic bomb over Hiroshima. Three days later, a second one destroyed the city of Nagasaki. On August 15, Japan declared its defeat. In 1947, the United States Strategic Bombing Survey explicitly stated that "the serious food problem not only undermined Japanese ability to prosecute the war, but also played a vital role in shaping Japan's political decision to surrender."[46] The report also emphasized that the dramatic shortage of food "was one of the several compelling arguments which determined the Japanese decision to surrender *even before the use of the atomic bomb* [emphasis added] and Russia's entry into the war."[47] The motivations underlying Japan's decision to surrender have been the subject of prolonged academic debate, that still remains ongoing. While an extensive discussion of this topic exceeds the scope of this work, it is essential to highlight the report's conclusions to illustrate how Operation Starvation effectively achieved its objectives by triggering a collapse

---

44 Ibidem.

45 United States Pacific Fleet and Pacific Ocean Areas, *Psychological Warfare, Part 2—Propaganda Material*, December 1944, Serial Number: 1006.

46 U.S.S.B.S., *The Japanese Wartime Standard of Living*, p. 108.

47 Ibidem, p. 109.

of the Japanese war economy and instilling fears among the populace regarding the impending catastrophe of widespread famines.[48]

1.4 *Takenoko no seikatsu* and the beginning of the Occupation

The unconditional surrender marked the cessation of the bombings, but it was still too early for the Japanese to speak of "postwar." Over two million people had lost their lives during the war and the economy of the country was completely annihilated. The extensive bombings had devastated urban and industrial infrastructures, and communication networks were seriously damaged. While the war had ended, Emperor Hirohito, in his surrender speech, emphasized that it was time for the Japanese to "endure the endurable and bear the unbearable." Yoshida Shigeru, the Prime Minister who dominated Japan's political scene in the postwar years until 1954, wrote that "the two-three years after the war were the darkest period in Japanese history."[49]

One of the most serious and compelling problem that Japanese people had to face in this moment was the scarcity of food. As a result of a number of factors such as the loss of its colonies, the destruction of its fleet, the shortage of fertilizers, Japan was unable to produce enough food for its population. People struggled to ride in crowded trains and flocked to the black market in the countryside. The train departing from Tokyo's Ikebukuro station to Saitama prefecture came to be known as the "potato train" (*imo densha*), due to the region's reputation for potato production.[50] Many people used to ask for food holidays (*shokuryō kyūka*) from work, in order to go to the black markets to buy foods. Even if it was illegal, the Metropolitan Police tacitly permitted these activities.[51] Due to the scarcity of goods and hyperinflation, bartering family possessions for basic sustenance, such as sweet potatoes or cabbages, became commonplace. By 1945, the practice of gradually getting rid of one's clothing, jewelry and other valuable items in exchange for food was referred to as *takenoko no seikatsu* or the "bamboo-shoot

---

48 Regarding the impact of the food problem on Japan's decision to end the war, see: Unno, Hiroshi, *Shokuryō mo daijōbu nari. Kaisen to shūsen no ketsudan to shokuryō* [English title: Starting and Ending the War (1941–1945). From the Perspective of Food Logistics], Tokyo: Nōrin tōkei shuppan, 2016.

49 Yoshida Shigeru, *Nihon wo ketteishita hyaku nen* [Japan's decisive century], Tokyo: Nihon Keizai Shinbunsha, 1967, p. 163.

50 Kishi Yasuhiko, *Shoku to nō no sengoshi* [The History of Food and Agriculture in the Postwar], Tokyo: Nihon keizai shinbunsha, 1996, p. 12.

51 Ibid., p. 11.

life," analogous to bamboo shoots being peeled away in layers, akin onions. The prevailing sense of despair and deprivation was vividly illustrated by the emergence of a new type of criminal, known as the "vegetable thieves," who stole food from fields at night before it could be harvested.[52]

If, during the war, attempts were made to offset the shortage of rice and other foods by carefully managing available resources and incorporating unconventional ingredients like weeds and potato peels,[53] the severe conditions following the defeat intensified the need for new solutions. In Osaka, scientists shifted their focus from developing military weapons to researching emergency diets. They taught people how to make the best of acorns, peanut shells, and even sawdust, fermenting the latter into a powder to supplement flour in bread or dumplings making. They were also taught that they could eat the flesh of mice and rats as well, if cooked properly. Blossoms, roses, and silkworm cocoons, once part of a pleasing and productive natural environment, became part of the new diet.[54] The situation was so desperate that there was also a sharp increase in the number of abandoned foundling babies left in the streets, as their parents struggled to find sufficient nourishment for them. In November 1946, the police reported discovering the bodies of 11 infants aged between one and three months in alleys and vacant lots.[55]

In September 1945, the occupiers began to arrive in Tokyo. Officially, the Occupation was managed collectively by the Allied Powers. During the Moscow conference in December 1945, which involved the Foreign Ministers from the United States, the United Kingdom and the Soviet Union, it was decided to establish two main bodies that would help manage the operations of the Occupation, the Allied Council for Japan, based in Tokyo, and the Far Eastern Commission (F.E..C), based in Washington. The Allied Council for Japan, representing the United States, British Commonwealth, China, and the Soviet Union, was created to offer advise to the Supreme Commander for the Allied Powers (S.C.A.P.) regarding the administration of Japan's Occupation. The F.E.C., consisting of 11 members, was entrusted with formulating Occupation policies, reviewing S.C.A.P.'s actions, and ensuring the implementation of the terms of the Potsdam

52 Dower, *Embracing Defeat*, p. 90; Collingham, *The Taste of War*, p. 309.

53 Saitō Minako, *Senka no reshipi. Taiheiyō sensō-ka no shoku wo shiru* [Recipes from the War: Understanding Food During the Pacific War]. Tokyo: Iwanami Shoten, 2015.

54 Christopher Harding, *Japan Story: In Search of a Nation, 1850 to the Present*, London: Penguin U.K., 2018, pp. 222–23.

55 Anonymous, "Allies Orders Tokyo Government to Distribute Food and Clothing," *The New York Times*, 27 December 1945.

Declaration.[56] However, in reality, the United States wielded significant autonomy in its management of Japan's Occupation. General Douglas MacArthur, serving as the Supreme Commander, along with a predominantly American staff, primarily followed directives from the U.S. government and paid little attention to the decisions made by these bodies.

The initial American strategy in Japan was encapsulated in two words: demilitarize and democratize. According to Washington, the first aim of the Occupation policy was to dismantle all the vestiges of Japan's military, economic, and political structure. This approach is evident in the Basic Directive for Post-Surrender Military Government in Japan Proper (J.C.S. 1380/15), which stated,

> You [S.C.A.P.] will not assume any responsibility for the economic rehabilitation of Japan or the strengthening of the Japanese economy. You will make it clear to the Japanese people that:
> a. You assume no obligations to maintain, or have maintained, any particular standard of living in Japan, and
> b. That the standard of living will depend upon the thoroughness with which Japan rids itself of all militaristic ambitions, redirects the use of its human and natural resources wholly and solely for purposes of peaceful living, administers adequate economic and financial controls, and cooperates with the occupying forces and the governments they represent.[57]

It was established that it was the responsibility of the Japanese authorities to avoid "acute economic distress" and "to provide goods and services to meet the needs of the occupying forces to the extent that this can be effected without causing starvation, widespread disease and acute physical distress."[58]

On September 21, during an interview with Hugh Baillie, the President of the United Press, General MacArthur was asked about Japan's potential to regain its status as a world power. In response, he emphatically stated that: "Never again, when the terms of the surrender have been accomplished"; he also declared, "*Her food supplies are scarce*, and she faces conditions in this emergency that may well become catastrophic. *Her punishment for her sins, which is just beginning, will be long and bitter* [emphasis added]," and also added, "Japan can expect no relief, *no*

[56] The members were: United States, Soviet Union, Republic of China, United Kingdom, the Philippines, Australia, New Zealand, Canada, France, India, and the Netherlands.

[57] Basic Initial Post-Surrender Directive to Supreme Commander for the Allied Powers for the Occupation and Control of Japan (JCS1380/15). Full text available at: <https://www.ndl.go.jp/constitution/e/shiryo/01/036/036tx.html> [accessed 21 March 2021].

[58] Ibidem.

food, clothing, or supplies from the Allied Powers this winter [emphasis added]."[59] While it remains uncertain to what extent MacArthur's statements represented his personal views or those of Washington, the interview undoubtedly revealed the prevailing American intentions concerning Japan's Occupation.

The problem of food shortage immediately emerged as one of the primary concerns confronting both the Occupation Forces and the Japanese government. Initially, the estimated rice production for 1945 stood at over 8.3 million metric tons as of August 31, 1945—a level deemed nearly normal. However, a devastating typhoon struck Western Japan in mid-September, causing extensive damage to rural areas and compelling authorities to drastically revise their prediction downwards at approximately 6.8 million metric tons.[60] The dearth of fertilizers further compounded the problem, as production plummeted from an annual average of 1,143,000 metric tons between 1937 and 1940 to a mere 300,000 metric tons in 1945.[61] Additionally, the repatriation of millions of Japanese citizens placed even greater strain on the already tenuous food supply. As a result, the Ministry of Agriculture and Forestry estimated that Japan could produce only 1,375 calories per person per day.[62] The S.C.A.P. itself admitted that there was "no hope of increasing the ration of rice ration" during the following year.[63]

Driven by the need to avoid an acute food shortage in the short period, on September 18, 1945, the newly formed government led by Prince Higashikuni Naruhiko submitted a document called "Request for support from the Allied Forces regarding food imports (Shokuryō yunyū ni kanshi rengō gun gawa ni taishi shien hōyōsei no ken)" to G.H.Q. This document outlined Japan's urgent

---

59 United States Congress, *Congressional Record: Proceedings and Debates of the 79th Congress—First Session*, vol. 91, Part 7, Washington D.C.: U.S. Government Printing Office, 1945, p. 8889.

60 Supreme Commander for the Allied Powers, *Summation No. 1 of Non-Military Activities in Japan and Korea*, Tokyo: General Headquarters Supreme Commander for the Allied Powers, September-October 1945, p. 44.

61 Ibidem, p. 45.

62 Steven J. Fuchs, "Feeding the Japanese: Food Policy, Land Reform, and Japan's Economic Recovery," in *Democracy in Occupied Japan: The U.S. Occupation and Japanese Politics and Society*, ed. Mark E. Di Caprio and Sugita Yoneyuki, London: Routledge, 2007, p. 28.

63 The national daily ration of 2.3 *go*, or about 330 grams of staple foods, was reduced to 2.1 *go*, or about 297 grams, in July 1945. The ration was originally intended in rice, but, due to the shortage of this commodity, the ration was supplemented by wheat, barley, and sweet potatoes. S.C.A.P., *Summation No. 1 of Non-Military Activities in Japan and Korea*, p. 108.

need to import approximately three million tons of food to avert famine.[64] In particular, the government asked the permission to import: 1.2 million tons of rice (500 thousand tons from Korea, 200 from Taiwan and 500 from other sources); 500 thousand tons of soybeans from Manchuria; and 1.3 million tons of wheat from the United States, Canada and Australia.[65] However, Colonel Raymond C. Kramer, head of the Economic and Scientific Division of S.C.A.P., expressed opposition to the Japanese government's requests. Despite recognizing the country's profound distress, he strongly criticized Japan's lack of efforts in mitigating widespread starvation and affirmed that the extent of American help depended upon "Japan's willingness to help herself."[66] Moreover, Colonel Kramer argued that even if food imports were permitted, Japan lacked the means to transport the necessary quantities due to the severe depletion of its fleet and shortages of gasoline and nets.[67] The Japanese government also sought permission from the Allies to borrow their ships for food imports, but MacArthur, though taking a "sympathetic view," refused as even the Allies did not have enough ships to permit the resumption of trade on such a scale and they needed them to transport American divisions home and furnish aids to other Allied countries like China and Philippines.[68] Besides, the Japanese plan lacked clarity on payment arrangements, as they did not furnish statistics on exportable goods to be used for paying for food and clothing imports.[69] S.C.A.P. clarified that no request to import commodities would be approved without an accompanying plan to establish credits trough exports.[70]

The Subcommittee for the Far East of the State-War-Navy Coordinating Committee (S.W.N.C.C.) in Washington, a federal committee created in 1944 to address the political and military issues related to the occupation of Germany and Japan, reiterated that "[r]elief should be accomplished with a maximum of Japanese responsibility and a minimum of imports" and that "the Supreme

64 J.A.C.A.R. (Japan Center for Asian Historical Records), "Shokuryō ni kanshi rengōgun gawa ni taishi shien hō yōsei no ken [Request for support from the Allies regarding food imports]," Ref.A17112842400, National Archives of Japan, URL: <https://www.digital.archives.go.jp/img.pdf/3761195> [accessed 20 March 2021].

65 Ibidem.

66 Anonymous, "Japan Is Lax in Readying for Winter," *Pacific Stars and Stripes*, 3 October 1945.

67 Burton Crane, "Need Will Govern Imports by Japan," *The New York Times*, 26 November 1945.

68 Anonymous, "Japan Seeks Ships for Food Imports," *The New York Times*, 8 November 1945.

69 George E. Jones, "Allies Will Split Japanese Trust," *The New York Times*, 16 October 1945.

70 S.C.A.P., SCAPIN-110, "Import of Essential Commodities," 9 October 1945, URL: <https://dl.ndl.go.jp/info:ndljp/pid/9885173> [accessed 21 March 2021]

Commander should assume no obligations to maintain [...] any particular standard of living in Japan" and "should not provide relief supplies unless it can be shown that a balanced diet of 1,500 calories of rationed and non-rationed items is not available for urban consumption."[71]

With the aim of preventing Japan from suffering severe hunger without providing direct food aid, the Subcommittee also proposed various measures:

- First of all, with the exception of tea and mulberry plantations, farmlands should have been converted to food crops to increase production;[72]
- Supplies of fertilizers and of essential spare parts and raw materials for fertilizers plants should have been authorized to the extent required to meet deficiencies in local supplies but it was specified that imports should have been procured as far as possible from sources in proximity to Japan;
- The production of farm tools and implements in short supply should have been encouraged;
- Stock from the Army remount depots should have been distributed to areas most dependent on animal power;
- Measures such as price control, rationing and control of black markets should have been strengthened in order to insure an equitable distribution of farm produce;
- The production of synthetic vitamins should have been encouraged to supplement the inadequacies of diet.[73]

That the food problem was not the highest priority for the Occupation Forces was clear to the Japanese authorities as well. Shortly after assuming office as Prime Minister on October 9, 1945, Shidehara Kijūrō announced eight important policies of his Cabinet, the first two being: (1) the establishment of democratic political policies and programs; and (2) the solution of the food problem. It is significant that "democracy" was placed ahead of the solution of the food problem. As George

71 "Report by the State-War-Navy Coordinating Subcommittee for the Far East," in *Foreign Relations of the United States: Diplomatic Papers 1945, The British Commonwealth, The Far East*, Volume VI, Document 513, Tokyo, October 1, 1945. URL: <https://history.state.gov/historicaldocuments/frus1945v06/d513> [accessed 21 March 2021]

72 Mulberry (for the production of silk), tea and the pyrethrum (a plant member of the chrysanthemum family, used for the production of insecticides) were considered the main industrial crops to exploit to increase Japanese export. See: S.C.A.P., *Summation No. 1*, p. 46.

73 "Report by the State-War-Navy Coordinating Subcommittee for the Far East," in *Foreign Relations of the United States: Diplomatic Papers 1945, The British Commonwealth, The Far East*, Volume VI, Document 530, Tokyo, October 11, 1945. URL: <https://history.state.gov/historicaldocuments/frus1945v06/d530> [accessed 21 March 2021].

C. Atcheson Jr., adviser to S.C.A.P., observed, in the minds of the vast majority of Japanese people, the food problem was paramount, almost to exclusion of the other problems.[74] The fact that the Japanese government put greater emphasis on the establishment of "democracy" clearly reveals the will to appease the Allies and accommodate the objectives of the Occupation.

However, they had immediately to face the negative consequences of the bad harvest and of the lower food production. While the Public Health and Welfare Section (P.H.W.) of the S.C.A.P. stated that there was no evidence whatsoever of deaths from starvation and that the problem of food was an exaggeration of the press "without any factual background,"[75] it was evident that most Japanese relied on the black market or other irregular channels to obtain approximately half of their food supply. Statistics revealed also that, during the first years after the war, the Engel's coefficient, indicating the proportion of income spent on food, exceeded 60 percent of Japanese income. This situation hindered the country's economic recovery and led to heightened social tensions.[76]

On October 16, the first "hunger march" took place in Tokyo. Several hundred persons charged into the Premier's official residence, then surged down the streets to take up a position in front of General Douglas MacArthur's headquarters, where they presented petitions urging the Allied Commander to "save us from starvation." They called for the resignation of the Prime Minister, the Cabinet, the Diet and other political leaders; the suicide of war leaders and confiscation of their property and all war profits; distribution of the surplus foodstuffs from military warehouses.[77] On November 15, in the village of Toyono, in Hokkaidō, thousands of people broke into a flour mill, assaulted and robbed farm carts carrying grains and ran away with whatever they could find. Over one thousand rioters attacked stores of food at three other places on the day after. In one case the rioters left behind them money equivalent to the Government's fixed price. Apparently, this was a protest against the government's policy of setting a price which was so low that the producers preferred to retain their food.[78]

---

[74] "The Acting Political Adviser (Atcheson) in Japan to the Secretary of State," in *Foreign Relations of the United States: Diplomatic Papers 1945, The British Commonwealth, The Far East*, Volume VI, Document 560, Tokyo, October 24, 1945. URL: <https://history.state.gov/historicaldocuments/frus1945v06/d560> [accessed 22 March 2021]

[75] S.C.A.P., *Summation No. 3 of Non-Military Activities in Japan and Korea*, Tokyo: General Headquarters Supreme Commander for the Allied Powers, December 1945, p. 151.

[76] Yoshida, *Nihon wo ketteishita hyaku nen*, p. 76.

[77] George E. Jones, "Hunger Marchers Call on M'Arthur," *The New York Times*, 18 October 1945.

[78] Lindsay Parrott, "Major Food Riots Stir North Japan," *The New York Times*, 17 November 1945.

On November 23, two hundred people broke into an American Army warehouse at Sendai, seizing food and other goods. It was the first outbreak to be reported in Japan involving American supplies. The police arrested 180 protesters, including 94 women.[79]

In response to Washington's concerns regarding the social stability of Japan, George C. Atcheson Jr. wrote directly to President Truman and reassured him that food riots had been a recurring aspect of Japanese history and very few Americans were preoccupied that such disturbances could create a military problem. Nonetheless, Atcheson also declared that it was fundamental to provide sufficient relief supplies in order to "maintain sufficiently solid ground" for advancing the Occupation's political objectives.[80] In another letter sent on January 4, 1946, he expressed even more clearly and vividly the imperative of maintaining a comprehensive vision in the management of the Occupation:

> One fundamental factor that we must not lose sight of in dealing with Japan is the inescapable relationship of politics and economics. In the dire straits of the Japanese people today, political development depends upon solution of the problems of food, shelter and clothing. Today political education and experience in the practice of democracy are impeded by the preoccupation of the people with their economic distress. Left unsolved, these problems may encourage extremism of types we do not desire.[81]

1.5 The Allies' contrasts over Japan's food imports

George Atcheson Jr.'s concerns about Japan's dire food situation were fully shared by General MacArthur as well. The urgency of addressing the looming food crisis became a focal point on December 20, 1945, when the Conference of the Presidents of Prefectural Legislatures of Japan convened to deliberate on the precarious state of the nation's food supply. They appealed to the Central Liaison Office (C.L.O.), the official communication channel between the General

---

79 The Associated Press, "U.S. Army Supplies Raided by Japanese Food Rioters," *The New York Times*, 24 November 1945.

80 "The Acting Political Adviser (Atcheson) in Japan to the Secretary of State," in *Foreign Relations of the United States: Diplomatic Papers 1945, The British Commonwealth, The Far East*, Volume VI, Document 593, Tokyo, November 5, 1945. URL: <https://history.state.gov/historicaldocuments/frus1945v06/d593> [accessed 21 March 2021]

81 "The Acting Political Adviser (Atcheson) in Japan to the Secretary of State," in *Foreign Relations of the United States, 1946, The Far East*, Volume VIII, Document 110, Tokyo, January 4, 1946. URL: <https://history.state.gov/historicaldocuments/frus1946v08/d110> [accessed 22 March 2021]

Headquarters and the Japanese government, to request S.C.A.P.'s authorization for the immediate importation of three million tons of food. This measure, they argued, was critical to prevent a catastrophic depletion of food stocks, projected to run dry by May 1946.[82]

MacArthur, acutely aware of the broader implications of such a crisis, recognized that famine on a national scale would not only result in widespread suffering but could also ignite a cascade of destabilizing consequences. The spread of disease, civil unrest, and an upsurge in illicit activities, such as the smuggling of rice reported in January 1946 from Korea,[83] were not merely hypothetical scenarios but real threats to the fragile social and political fabric of occupied Japan. These developments, if left unchecked, would severely undermine the foundational objectives of the Allied Occupation, jeopardizing efforts to rebuild Japan as a stable and cooperative member of the international community. Confronted with the Japanese government's struggles to boost national production in the short term and the deteriorating social conditions, MacArthur, like Atcheson, began to believe that some form of direct aid was necessary. Importing a substantial quantity of food was the initial step, but this required permission and cooperation from the U.S. government. In early January 1946, MacArthur sent a report to Washington detailing Japan's grim food situation, echoing the preoccupations of the Presidents of Prefectural Legislatures and emphasizing the need for Japan to import food. "Thirty million people are threatened by starvation," he warned; "unrest and disease will endanger the work the Occupation."[84] The urgency of his plea was underscored by projections from S.C.A.P., which indicated that by May or June of that year, Japan's food reserves would dwindle to the barest minimum necessary to sustain the population. Without immediate intervention, widespread starvation would become inevitable. The Office of S.C.A.P. echoed these dire warning and reported: "it has been determined that there will not be sufficient food to carry over the population, from about May until the next harvest, on a minimum essential caloric intake."[85] Meanwhile, the Chief of the Price Control

---

[82] J.A.C.A.R., "Shokuryō kiki dakai ni kansuru konsei no ken [Solicitation for overcoming the food crisis]," Ref.B20010397500, E'.2.0.0.9-7-2-1, Diplomatic Archives of the Ministry of Foreign Affairs of Japan, URL: <https://www.jacar.archives.go.jp/das/image/B20010397500> [accessed 22 March 2021]

[83] S.C.A.P., SCAPIN-660, "Violations of S.C.A.P. directive no. 3," 28 January 1946, URL: <https://dl.ndl.go.jp/info:ndljp/pid/9885730> [accessed 23 March 2021]

[84] E. J. Lewe van Aduard, *Japan: From Surrender to Peace*, Dordrecht: Springer Science & Business Media, 2013, p. 59.

[85] George Hubbard Blakeslee, *The Far Eastern Commission: A Study in International Cooperation: 1945 to 1952*, Washington D.C., Department of State, 1953, pp. 25–6.

and Rationing Division of S.C.A.P. reinforced the need for immediate action, declaring that food imports were not only essential but also urgent, emphasizing "the sooner the better."[86] The general perception was that within S.C.A.P., the food problem had become the utmost priority and that neglecting it could have led to catastrophic human consequences.

Despite the appeals, the S.C.A.P. was informed that the critical global food shortage and the threat of starvation in other liberated areas of Europe made it impossible to allocate any food imports to Japan. At this juncture, the Supreme Commander made it clear that it was not only about the fate of the Japanese people and the prevention of unrest and disease, as specified in the post-surrender directive, but he also feared that such a famine would seriously impede the constructive progress made in democratizing Japan, a primary objective for the United States. When the Appropriations Committee of the United States House of Representatives asked him to justify such a high expenditure to feed their former enemies, he frankly said:

> To cut off Japan's relief supplies in this situation would cause starvation to endless Japanese [...] and starvation breeds mass unrest, disorder and violence. Give me bread or give me bullets.[87]

Notwithstanding Atcheson's and MacArthur's stances, Washington remained hesitant to comply with their request. In a meeting on February 26, the Secretaries of the Departments of State, War, Commerce and Agriculture considered S.C.A.P.'s requirements but wanted to verify the accuracy of the conditions in Japan compared to other areas.[88] They decided to send Colonel Raymond L. Harrison to Tokyo in March 1946 to survey Japan's food needs. The Harrison Mission supported S.C.A.P.'s view concerning the urgent need of allocating food imports for Japan. In particular, in a letter sent to the Secretary of State and the Secretary of Agriculture about the findings of his group, Colonel Harrison pointed out:

> While the investigations made here to date have not disclosed any starvation areas, they clearly indicate that unless [substantial] imports of food [...] are actually in Japan beginning not later than 1 May 1946, there will arise in Japan during the period May through September 1946 a very critical shortage of food which will gradually spread throughout

---

[86] Ibid., p. 26.

[87] Douglas MacArthur, *Reminiscences*, New York: McGraw-Hill, 1964, p. 307.

[88] "Incoming Message from Washington [Department of War] to S.C.A.P.," 28 February 1946, in National Archives and Records Service, "G.H.Q./S.C.A.P. Records," Box no. ESD-2, Folder Title: "World Food Crisis," File #21, Feb. 1946–Apr.1946.

> the country and undoubtedly result in a decrease in indigenous food and industrial production, disease and unrest with the possibility of disorder of increasing severity.[89]

For these reasons, he recommended that 200,000 tons of grain a month be shipped immediately from March to June 1946.[90] On March 10, Major W.H. Leonard, Chief of the Agriculture Division of Allied Headquarters' Natural Resources Section, also declared that May 1 was "the absolute deadline before which substantial imports must arrive" because, beyond that date, the reality of an almost complete depletion of the nation's food reserves would be unavoidable, plunging Japan into a crisis of unprecedented magnitude.[91] In light of the acute world food shortage, the United States agreed to provide Japan with 150,000 tons of cereal per month from April to June, instead of the 200,000 originally requested by MacArthur. Yet, by April, only 48,000 tons of food had been shipped to Japan.[92]

Simultaneously, the Japanese government attempted to bolster domestic food production and improve distribution, though some measures fell short of addressing the severity of the crisis. On February 17, 1946, the government issued the Emergency Imperial Food Ordinance, introducing stringent legal measures aimed at increasing crop collections. This law authorized expropriation and punitive actions against those farmers who failed to deliver their food quotas. It was believed that enhancing crop collection from farmers would result in an increase in the distribution of staple food in urban areas, subsequently leading to a decrease in the availability of food for the black market. However, the Allied authorities remained skeptical of the ordinance's efficacy. The Natural Resources Section (N.R.S.) thought that these coercive policies might indirectly limit the food available to rural population and sow deep resentment among farmers.[93]

Faced with the dual challenge of insufficient food imports from the United States and the inadequacy of measures implemented by the Japanese government, General MacArthur urgently turned his attention to alternative sources in Asia. Yet, his efforts proved fruitless. The first strategy involved exploring the possibility of importing rice from Korea, a proposal that aligned with suggestions from the

---

89 Bowen C. Dees, *The Allied Occupation and Japan's Economic Miracle: Building the Foundations of Japanese Science and Technology 1945–52*, London: Routledge, 1997, p. 88.

90 Blakeslee, *The Far Eastern Commission*, p. 178.

91 Burton Crane, "May 1 Is Deadline for Food to Japan," *The New York Times*, 11 March 1946.

92 Blakeslee, *The Far Eastern Commission*, p. 178.

93 Natural Resources Section to Economic and Scientific Section, "Food Crisis in Japan Ref. W98710," Serial Number 21-13, 1 March 1946, in National Archives and Records Service, "G.H.Q./S.C.A.P. Records," Box no. ESD-2, Folder Title: "World Food Crisis," File #21, Febr. 1946–Apr.1946.

Presidents of Prefectural Legislatures.[94] At that time, Korea was enjoying the benefits of an exceptionally good harvest in 1945, resulting in a substantial surplus of rice. However, Korea's coal industry—vital for its recovery—was faltering due to severe shortages. A trade arrangement between the two nations, in which Japan would export coal to Korea in exchange for rice, appeared not only practical but mutually advantageous. However, geopolitical realities soon dashed these hopes. The Soviet Union, which had significant influence over Manchuria, redirected food exports from the region—upon which Korea depended for its millet supply—toward Siberia. This disruption in Korea's food imports created domestic uncertainty, prompting Korean farmers to hoard their surplus rice rather than release it for export. The result was a complete breakdown of negotiations, with no rice becoming available for Japan.[95] MacArthur's search for alternatives extended to Thailand, the only Asian country besides Korea known to possess significant rice surpluses. However, this option also proved unviable, as the United Nations Relief and Rehabilitation Agency (U.N.R.R.A.) purchased all Thai surplus stocks to address the broader regional needs of postwar Asia.[96]

With no clear action from Washington after more than five months of protracted negotiations, and no viable option for food imports from neighboring countries, S.C.A.P. took matters into its own hands and authorized the distribution of substantial quantities of surplus food stocks, including U.S. Army reserves, Japanese military surpluses, and supplies managed by the Japanese government. In a particular significant move, MacArthur ordered the transfer of approximately 3,500 tons of wheat flour—originally shipped from the Philippines and deemed "surpluses" to the needs of Allied Forces in the Pacific—to the Japanese government.[97] Although the ration provided still fell below minimum dietary requirements, these measures ensured a minimum subsistence level.[98]

However, more than the obstacles posed by Washington, the greatest resistance that McArthur and Atcheson had to face was within the Allied Forces themselves. In fact, in addition to the difficulty of identifying a proper source of food supply abroad, McArthur and Atcheson had to also face a strong opposition from some members of the F.E.C., who were against any initiative that might be interpreted as affording Japan preferential treatment over other Asian nations, whose precarious

---

[94] J.A.C.A.R., "Shokuryō kiki dakai ni kansuru konsei no ken."

[95] Burton Crane, "Allies in Dilemma on Feeding Japan," *The New York Times*, 10 January 1946.

[96] Burton Crane, "No Food in Sight for Hungry Japan," *The New York Times*, 16 March 1946.

[97] Lindesay Parrott, "Japan Gets Bread from Allied Flour," *The New York Times*, 15 March 1946; S.C.A.P., SCAPIN-818, "Delivery of Wheat to Japan," 15 March 1946, URL: <https://dl.ndl.go.jp/info:ndljp/pid/9885893> [accessed 24 March 2021].

[98] Fuchs, "Feeding the Japanese," p. 33.

circumstances were a direct consequence of Japan's brutal actions during the war. The issue of food policy was raised in the F.E.C. by Sir Carl Berendsen, the New Zealand representative, supported by the representatives of India and Philippines. He stated that the food shortage was a worldwide problem and the inadequacy of food supplies was "brought about by the criminal attack on civilization made by the Axis Powers" and then the Japanese were "themselves, in substantial part, the cause of the famine that menaces the world."[99] Although several months had passed since the surrender, Sir Berendsen still regarded Japan as an enemy unworthy of mercy, insisting that even the civilian population deserved no special consideration. For this reason, he did not refrain from criticizing the United States as well. He affirmed that it seemed wrong that Japan was receiving special American food aids when China and India, the victims of Japan's wartime aggression, were suffering even a more severe food crisis.

On April 25, the F.E.C. approved a document reaffirming the intention to continue treating Japan as an enemy nation, not deserving any preferential treatment, even when faced with clear difficulties for the civilian population. The document, titled "Food Supply for Japan," delineated two essential points:

1. strict prohibition of any imports that might confer the Japanese with priority or preferential treatment over the requirements of the people of any Allied Power or liberated area;
2. a formal request from the Far East Commission to the United States Government, urging an immediate review of the food import program for Japan in accordance with the aforementioned policy.[100]

General MacCoy, the U.S. representative, expressed sympathy for Sir Berendsen's apprehensions regarding the dire global food situation but raised doubts about the F.E.C.'s jurisdiction over food allocation for Japan, and defended the U.S. Government's decision.[101] Nonetheless, the tensions among the Allies on this issue persisted. In a news conference, Sir Ben Smith, the British Food Minister, stated that he would advocate within the Combined Food Boards for a policy prioritizing wartime Allies' access to the limited global food supplies, placing former enemy nations "at the end of the queue."[102]

---

[99] Blakeslee, *The Far Eastern Commission*, pp. 176–77.

[100] The Far Eastern Commission, "Activities of the Far Eastern Commission: Report by the Secretary General, February 26, 1946–July 10, 1947," Washington D.C.: Department of State, Division of Publications, Office of Public Affairs, 1947, Appendix 35, p. 90.

[101] Blakeslee, *The Far Eastern Commission*, p. 179.

[102] Bess Furman, "Feed Allies First, Ben Smith Urges," *The New York Times*, 14 March 1946.

The issue of food aid to Japan was brought to the forefront of the Allied Council upon the request of General William Macmahon Ball, the representative of the British Commonwealth. A staunch advocate of the "hard line" policy against Japan, Macmahon Ball had consistently pushed for measures that underscored accountability for wartime actions. His stance included advocating for the indictment of the Emperor as a war criminal, conducting extensive trials of Japanese war criminals, opposing the resumption of Japanese whaling in the Antarctic, and promoting the long-term demilitarization and supervision of Japan.[103] The purpose of his inquiry was to raise the question of whether the food provided by the United States to Japan should not be allocated to other countries instead. While acknowledging that starvation or even short rations were not "a proper penalty to impose to the Japanese people for the past crimes of their rulers" and that "everything should be done to see that the Japanese people have enough food," General Macmahon Ball also believed that the problem of food scarcity in Japan should be examined in the context of the general world shortage. He also stated that it was essential to understand whether Japan's needs were greater than the needs of other countries, implicitly criticizing U.S. decision to send food to Japan.[104]

In response to General Macmahon Ball's inquiry, George C. Atcheson Jr., who replaced General William F. Marquat as the Chairman of Council and representative of the United States in April 1946, responded with some irritation. He firmly clarified that the food was solely provided by the United States outside the scope of any international agreement, and therefore, it was solely subject to decision and determination by the United States Government. Atcheson emphatically stated that the food was sent to Japan "at some sacrifice by the American people" and there was no question that for the United States it was preferable "to send food to Japan that to have to send large numbers of additional soldiers."[105] While Atcheson advocated for an increase in food imports to Japan, he remained keenly aware of the political challenges posed by the Allies' criticism. These objections, if amplified, risked distorting public perceptions of S.C.A.P.'s efforts and jeopardizing its position, particularly among the American public, who harbored lingering resentment toward their wartime adversary. To counteract this, Atcheson stressed

[103] Christine de Matos, "Diplomacy Interrupted?: Macmahon Ball, Evatt and Labor's Policies in Occupied Japan," *Australian Journal of Politics & History*, Vol. 52, no. 2, 1 June 2006, p. 192.

[104] Allied Council for Japan, "Verbatim Minutes of the Third Meeting (Morning Session)," Tokyo, 30 April 1946, J.A.C.A.R., Ref. B19010254200, B'.1.1.0.2-2, p. 14. URL: <https://www.jacar.archives.go.jp/das/image/B19010254200> [accessed 25 March 2021]

[105] Ibid., pp. 15–6

the importance of ensuring transparency and providing accurate information to the press. He argued that publicizing the facts and context of Japan's dire situation was essential for garnering broader support and maintaining the credibility of S.C.A.P.'s mission during a highly delicate postwar period.[106]

1.6 The impact of *Food May Day*

The uncertain volume and timing of food imports, the difficulty in augmenting crop collection from farmers, and the numerous deficiencies in the distribution system engendered widespread discontent. The recently legalized Kyōsantō (Japanese Communist Party), whose members were released from jail by the S.C.A.P. in October 1945, endeavored to enhance its support by critiquing the Government and the Occupation Forces for their management of the food problem. After an eighteen-year incarceration, Tokuda Kyūchi assumed the role of General Secretary of the Kyōsantō from 1945 to 1952. He accused bureaucrats, capitalists, and even the Emperor of stealing food and all other necessities from the Japanese people.[107] Tokuda argued that the path to power in postwar Japan laid in addressing the widespread discontent caused by hunger. For him, control over food was the key to controlling the nation.[108] In his perspective, people should take over the control and management of food in storage and distribute the land of the Emperor, absentee landlords, and the rich farmers to the poor peasants.[109] To turn these ideas into action, the Kyōsantō established the *Shokuryō taisaku iinkai* (Committee on Food Countermeasures) in November 1945 and formulated a policy in February 1946, aiming to address the food crisis by forming a people's council that combined the labor unions, the peasant organizations, and the Citizen Food Management Committee (Shimin shokuryō kanri iinkai).[110] Building on this initiative, in April and May 1946, the Communist Party, alongside

---

106 "The Political Adviser (Atcheson) in Japan to the Secretary of State," in *Foreign Relations of the United States, 1946, The Far East*, Volume VIII, Document 190, Tokyo, April 26, 1946. URL: <https://history.state.gov/historicaldocuments/frus1946v08/d190> [accessed 24 March 2021]

107 Robert A. Scalapino, *The Japanese Communist Movement, 1920–1966*, Santa Monica: University of California Press, 1967, p. 50.

108 Theodore Cohen, *Remaking Japan: The American Occupation as New Deal*, ed. Herbert Passin, New York: Free Press, 1987, p. 145.

109 Scalapino, *The Japanese Communist Movement*, p. 50.

110 "Shokuryō Mēdē to Tennō Purakādo Jiken (2)—Matsushima Matsutarō-Shi Ni Kiku [The Food May Day and the Placard Incident (2)—Interview with Matsushima Matsutarō]," *Ohara Shakai Mondai Kenkyū-Sho Zasshi* [Journal of the Ōhara Institute for Social Research], n. 535, Tokyo: Hōsei University, June 2003, pp. 55–6.

labor unions, orchestrated mass demonstrations in Tokyo and other major cities, demanding larger food rations, higher wages, and the resignation of the Shidehara and later the Yoshida Cabinets. The pinnacle of these demonstrations was on May 19, 1946, known as the "Food May Day (shokuryō mēdē)," one of the largest mass protests that Tokyo has ever seen, where over 200,000 demonstrators gathered in front of the Imperial Palace, urging for a resolution of the food crisis through mechanisms of food distribution under popular control.[111] "Hunger stalked the streets, [...] and the red flags carried by demonstrators were everywhere to be seen," with these words Yoshida Shigeru vividly remembered the protest in his memoirs while striving to establish his first Cabinet in those days.[112] But there were not only the red flags. During the protests, straw-mat banners and placards with slogans like "Hatarakeru dake kuwasero [Let us eat enough that we can work]," "Kome wo yokose [Give us the rice]," "Kenpō yori meshi da [Food before a constitution]" or "Shokuryō jinmin kanri [Popular control of food]" were prominently displayed, highlighting the urgency for demonstrators to address the food crisis before pursuing any other reform.[113] The placard that more than any other attracted attention was the one held aloft by Matsushima Matsutarō, a factory worker and member of the Communist Party. His placard bore the inscription:

> Imperial Edict: The Kokutai has been preserved. I, the Emperor, have filled my belly, but you, common people, may starve to death! Signed: The Emperor.

This episode, later labeled as the "placard incident (purakādo jiken)," led to Matsushima being indicted for the crime of lese-majesté. In November 1946, he received an eight-month labor sentence, which was later revoked after a campaign advocating on his favor.[114]

The escalating popular discontent became a matter of serious concern for both S.C.A.P. and the Japanese government. The confirmation of a plot to assassinate MacArthur intensified the alarm, leaving the Japanese people feeling a sense of national humiliation and fearing potential repercussions on their food supplies from the United States.[115] On May 20, the day after the protests, MacArthur

---

111 Joe Moore, "Production Control: Workers' Control in Early Postwar Japan," *Bulletin of Concerned Asian Scholars* 17, n. 4, December 1985, p. 22.

112 Yoshida Shigeru, *The Yoshida Memoirs. The Story of Japan in Crisis*, trans. Yoshida Ken'ichi, London: William Heinemann Ltd., 1961, p. 279

113 "Shokuryō Mēdē to Tennō Purakādo Jiken (2)," p. 69.

114 Kishi, *Shoku to nō no sengoshi*, p. 16.

115 Anonymous, "MacArthur Plot Alarms Japanese; They See Possible Repercussions," *The New York Times*, 1 May 1946.

condemned the situation as "a menace not only to orderly government, but to the basic purposes and security of the Occupation itself."[116] Even Hirohito, in his first radio speech after he broadcast his order for Japan's surrender, decided to intervene in the matter, asking for a rapid solution of the food problem, which he considered "extremely serious and a source of worry." He urged the government to promptly implement appropriate measures and, at the same time, he pleaded that all Japanese citizens set aside personal interests and extended support to one other, sharing the burden of hardship and the distress together.[117] In his first statement as the new Prime Minister, Yoshida Shigeru pledged immediate action to solve the nation's critical food problem and announced a four-point program to be taken immediately to alleviate the crisis.[118]

Former American President Herbert Hoover and Dennis A. Fitzgerald, who would soon become the first Secretary-General of the International Emergency Food Council (I.E.F.C.) in June 1946, visited Japan in May 1946. The purpose of their visit was to lead a food mission on behalf of U.N.R.R.A., in order to conduct a worldwide survey, as requested by President Truman. General MacArthur reported to Hoover that Japan's food situation was the worst it had been in 30 years. Hoover and Fitzgerald reviewed the estimates of Japan's food import requirements and corroborated the recommendations of the Harrison Mission. After he had attested the precarious situation of Japan's food supply, Hoover sharply declared that:

> Japan must have food imports. Without them all Japan will be rationed little better than that which the Germans gave the inmates of Buchenwald and Belsen concentration camps. It is impossible to conceive that the American flag will fly where such conditions exist.[119]

As a result of the survey, Hoover recommended that 870,000 tons of food be imported to Japan by the end of September, with the largest quantities arriving during June and July when the food shortage was expected to be the most

---

116 Anonymous, "M'Arthur Puts Ban on Mass Agitation," *The New York Times*, 21 May 1946.
117 Hirohito's speech can be listened and read at: <https://www.kunaicho.go.jp/kunaicho/koho/taisenkankei/syokuryo/syokuryo.html> [accessed 25 March 2021]
118 The points of this program were: 1) the application of urgent measures to obtain from the farmers their quotas; 2) the compilation of a correct inventory of food present in stock but withheld from distribution; 3) the improvement of the supply system; 4) negotiations with the Allies for importing more food. See: Lindesay Parrot, "Japanese Premier Faces Food Issues," *The New York Times*, 23 May 1946.
119 Anonymous, "Hoover Bids Soviet Free Food Supply," *The New York Times*, 7 May 1946.

critical.[120] The evoked image of a concentration camp was too powerful to be opposed and the United States accepted to export about 600,000 tons of rice equivalents between May and September 1946. By the summer of 1946, one-quarter of the food consumed by the Japanese came from Allied sources, including U.S. Army reserve supplies in Japan.[121] It has been estimated that American food aid to Japan in 1946 saved around 11 million Japanese from starvation.[122]

Even MacMahon softened his position. While he still believed that there was enough food to avert starvation and the real problem derived from the incapacity of Japan's government to collect and distribute the food equally among the population, the mounting protests and the potential political unrest caused him to consider it necessary to provide aid to Japan's government "to avoid some of the political penalties of its own inefficiency."[123]

1.7 Navigating food scarcity: The Whaling Expedition in the Antarctic

Confronted with the difficulty of rapidly increasing agricultural production and food imports, S.C.A.P. identified fishing, more specifically whaling, as a viable solution to alleviate Japan's precarious food situation. Prior to the outbreak of the Pacific War, Japan ranked third among the whaling nations of the world, after Norway and the U.K. However, in the aftermath of the attack on Pearl Harbor in 1941, all whaling activities in high seas were suspended, and only coastal operations continued until they eventually ceased upon Japan's surrender in August 1945.[124] On September 22, 1945, in response to the critical shortage of protein-rich foods and edible oils during the initial phase of the Occupation, the S.C.A.P. authorized the resumption of fishing and whaling activities, albeit within

---

[120] S.C.A.P., *Summation No. 5 of Non-Military Activities in Japan and Korea*, Tokyo: General Headquarters Supreme Commander for the Allied Powers, May 1946, p. 179.

[121] In June 1946, the S.C.A.P. instructed Japanese government to send shipping to lift approximately 27,000 long tons of excess United States Army food stocks located in Okinawa, Manila, and San Fernando. See: G.H.Q., War Department, "Operations Report Japan," serial no. 312, n. C62121, 18 June 1946.

[122] Richard B. Finn, *Winners in Peace: MacArthur, Yoshida, and Postwar Japan*, Berkeley: University of California Press, 1992, p. 113–14.

[123] W. Macmahon Ball, "Reflections on Japan," *Pacific Affairs*, vol. 21, n. 1, 1948, p. 14.

[124] Supreme Commander for the Allied Powers–Natural Resources Section, *Fisheries Programs in Japan, 1945–51*, Report n. 152, Tokyo: General Headquarters, Supreme Commander for the Allied Powers. Natural Resources Section, 1951, pp. 15–16.

a specified coastal zone known as the MacArthur Line and in strict adherence to the International Whaling Regulations.[125]

Amidst worsening conditions following the Food May Day, during the Sixth Meeting on June 12, 1946, the Allied Council discussed on the possibility to extend the areas where Japanese fishing and whaling operations could be conducted. Fishery activities were recognized not only as a means to get food in a cheap and quick way but also as a driving force for economic revitalization. It was suggested that an increased fishing area would generate fish surpluses, which Japan could export in exchange for staple foodstuffs or other essential commodities.[126]

However, once again, Atcheson had to face skepticism and critical opposition from the Allies. General K. N. Derevyanko, member for the U.R.S.S., and Mr. Yorkson C. T. Shen, representing China, insisted that any measures should be only temporary.[127] Although General Dereyanko initially stated that "in principle" he had "no objections to permitting the Japanese to conduct fishing and whaling in the suggested area as a temporary arrangement,"[128] a subsequent letter to Atcheson contradicted this stance, revealing that his government had reversed its position and was now opposed to extending the existing Japanese fishing areas.[129] Atcheson perceived this reversal as "a continued Soviet desire to hamper the policies of the Occupation in Japan."[130]

Once again, MacArthur decided to take unilateral action without prior consultation with the Allied Governments and proceeded to extend the areas in the open seas where Japanese activities were permitted, on the basis either of his general executive authority or of the early fishing directive sent to S.C.A.P. by the American government in November 1945. The directive stated that S.C.A.P. possessed the authority to grant permission for Japan to engage in "deep sea fishing" anywhere in the open seas, except in waters deemed "near" the territories of the United States or of other Allies. However, the directive did not provide a clear definition of what constituted "near."[131] On the authority of this directive

---

125 Ibid., p. 16.

126 Allied Council for Japan, "Verbatim Minutes of the Sixth Meeting (Morning Session)," Tokyo, 12 June 1946, J.A.C.A.R., Ref. B19010270700, p. 1. URL: <https://www.jacar.archives.go.jp/das/image/B19010270700> [accessed 25 March 2021]

127 Ibid., pp. 3–12.

128 Ibid., p. 11.

129 "The Political Adviser (Atcheson) in Japan to the Secretary of State," in *Foreign Relations of the United States, 1946, The Far East*, Volume VIII, Document 219, Tokyo, July 1, 1946. URL: <https://history.state.gov/historicaldocuments/frus1946v08/d219> [accessed 24 March 2021].

130 Ibidem.

131 Blakeslee, *The Far Eastern Commission*, pp. 105–06.

and in response to a petition from the Japanese government, on August 6, 1946, S.C.A.P. official authorized whaling operations in the Antarctic for the 1946–47 season.[132] The principal motivations cited for the authorization were the acute food shortage in Japan and the world scarcity of oils and fats.[133]

Following the authorization, the Australian government, through its Embassy in Washington, sent an aide-mémoire to the Department of State, expressing strong concerns about MacArthur's proposal for a whaling expedition. These objections were rooted in past actions by Japanese ships that were perceived as prejudicial to Australian security, prior violations of International Whaling Regulations by the Japanese, and the inefficiency and wastefulness of Japanese whaling methods.[134] The Australian concerns were shared also by the United Kingdom and New Zealand, who were equally apprehensive about the potential revival of the Japanese fishing industry.[135] The United States replied that "the action of the Supreme Commander [...] was clearly within his authority" and confirmed the temporary and emergency nature of the expedition, which was closely linked to Japan's precarious food supply situation.[136] Washington also reassured its Allies that any future authorization for Japanese whaling expedition would be subject to prior consultations with the interested governments. It was also clarified that the expedition would remain under complete control of the Allies, and the ships involved would not fly the Japanese flag.[137] After these considerations and assurances from the U.S. government, the United Kingdom withdrew its protest against the Japanese whaling expedition, while Australia and New Zealand did not raise further objections.

S.C.A.P. granted authorization for the deployments of two fleets, for a total of 21 vessels, which also included two factory ships. Departing from Japan in mid-November 1946, these fleets were accompanied by two United States Army officers, acting as S.C.A.P. representatives, along with one Australian and one British observer. They operated in the Antarctic waters from December 16, 1946, to March 14, 1947, and were able to take 1,164 whales.[138]

---

132 S.C.A.P., SCAPIN-1103, "Japanese Whaling Operations in the Antarctic," 6 October 1946, URL: <https://dl.ndl.go.jp/info:ndljp/pid/9886213> [accessed 25 March 2021].

133 "The Political Adviser (Atcheson) in Japan to the Secretary of State," in *Foreign Relations of the United States, 1947, The Far East*, Volume VI, Document 229, Tokyo, June 26, 1947, URL: <https://history.state.gov/historicaldocuments/frus1947v06/d229> [accessed 25 March 2021].

134 Ibidem.

135 Blakeslee, *The Far Eastern Commission*, p. 105.

136 Ibid., p. 106.

137 Ibidem.

138 S.C.A.P., *Fisheries Programs in Japan, 1945–51*, p. 16.

Due to the success of the first post-hostilities expedition and the ongoing demand for whale products, two-fleet expeditions set sail each year thereafter until the conclusion of the Peace Treaty in 1951, despite facing continuous opposition from other F.E.C. members.[139] According to S.C.A.P., the whaling expeditions conducted between 1946 and 1951 proved highly effective, yielding approximately 251,000 metric tons of oil and food products that significantly mitigated the food crisis in Japan. Additionally, these expeditions played a crucial role in transforming the attitude of both government officials and major whaling companies toward international cooperation, representing a significant shift from the prewar era. This change ultimately culminated in Japan's adherence to the International Convention on the Regulation of Whaling in 1951.[140]

1.8 Conclusion

The significance of food in wartime economies cannot be overstated, as it occupies a position of utmost importance for governments engaged in conflicts. Ensuring stable food supplies is vital, not only for maintaining the physical strength of the military but also for preserving civilian morale. By 1945, Japan found itself in a preoccupying situation with its food supply. Prewar reliance on colonies and occupied territories for food became increasingly difficult as the war prolonged. Additionally, domestic production declined due to labor and resource reallocation for military purposes. The devastating impact of Operation Starvation further exacerbated Japan's food crisis, cutting off any external food supply and creating fears of a mass famine among the population. Even after Japan's surrender in 1945, the food situation remained critical, but as previously discussed, Japan was denied any preferential treatment, with the responsibility for resolving the crisis placed squarely on the shoulders of the Japanese government. Although Tokyo initially sought minimal imports to mitigate the crisis, S.C.A.P. dismissed their appeals, adhering to the principle that Japan's recovery should be self-reliant.

Yet, as we have sought to demonstrate, the sheer gravity of the food crisis soon exposed divisions within the occupying forces themselves. General Douglas MacArthur and George C. Atcheson Jr. were among the first to acknowledge that resolving the food issue was essential to achieving the core objectives of the Occupation, including stabilization and democratization. However, their calls for assistance met resistance on multiple fronts. In Washington, policymakers hesitated, wary of the political and financial implications of intervention. Meanwhile,

---

139 Blakeslee, *The Far Eastern Commission*, 106–114.

140 S.C.A.P., *Fisheries Programs in Japan, 1945–51*, p. 16–18.

the United Kingdom and Commonwealth nations adopted a hardline stance, opposing any aid that might be seen as undue leniency toward their former adversary. Similarly, China and the Soviet Union exhibited strong resistance, particularly when discussions arose about resuming Japan's open-sea whaling activities, which symbolized broader resource exploitation.

In this chapter, we have sought to illustrate how the worsening conditions in Japan, coupled with the looming threat of social unrest, ultimately compelled the United States Government to reconsider its position. Washington's reluctant agreement to MacArthur and Atcheson's proposals marked a pivotal shift in its approach toward Japan, motivated by the realization that addressing food insecurity was indispensable for stabilizing the nation and achieving Occupation goals. Meanwhile, other Allied Powers, though initially resolute in their hardline positions, began to soften under the dual pressures of American unilateral action and the growing recognition of the urgent need to quell social tensions. The gradual alignment of Allied perspectives, driven by pragmatism and necessity, underscores the complex interplay of diplomacy, strategy, and humanitarian concern during this critical chapter in Japan's postwar reconstruction.

CHAPTER 2

Helping Japan: Food aid, militarization, and the economic recovery

2.1 Introduction

The period extending from the beginning of Operation Starvation in March 1945 to the summer of 1946 represents a particularly traumatic chapter in Japan's recent history in relation to nutrition and food security. As outlined in the first chapter, the conclusion of the war did not immediately herald relief; rather, it precipitated a further decline in living conditions. This period saw the Japanese government grappling with the imminent threat of widespread starvation, driven by the inability to access international trade networks and the absence of sufficient food aid. As previously discussed, the first half of 1946 was marked by significant divisions within the Occupation Forces regarding the most effective strategy to mitigate the food crisis. However, escalating social unrest and the deepening severity of the situation necessitated a shift in the Occupation's approach. The whaling expeditions authorized by S.C.A.P. in the summer of 1946 may have been the first attempt by the occupying forces to address Japan's food crisis by allowing the country to look for food resources beyond its own borders. Nevertheless, the acute food emergency was ultimately alleviated through the arrival of American food aid, which played a critical role in stabilizing conditions and laying the groundwork for a sustainable resolution to the crisis.

This chapter examines the pivotal role of American food assistance in addressing Japan's postwar "food problem." It explores key programs implemented during the Occupation—such as L.A.R.A. (Licensed Agencies for Relief in Asia), G.A.R.I.O.A. (Government Aid and Relief in Occupied Areas), and E.R.O.A. (Economic Rehabilitation in Occupied Areas)—as well as initiatives introduced in the 1950s following the restoration of Japan's sovereignty through the San Francisco Peace Treaty, including M.S.A. (Mutual Security Assistance) and P.L.480 ("Food for Peace" Public Law 480). These programs were driven by three distinct imperatives: first, a humanitarian need to address the pressing food crisis; second, an economic necessity that involved managing American agricultural surpluses and reviving industrial production in Japan; and third, a diplomatic objective to prevent the spread of Communist ideologies in Japan and ensure its alignment with the free world. The three objectives were closely interconnected. A 1948 C.I.A. document emphasized that Japan's defeat in World War II had created a power vacuum in East Asia. Both the United States and the Soviet Union recognized that controlling Japan was crucial for exerting influence in the region due to its strategic location and potential manpower resources. From the U.S. perspective, Japan's economic recovery was considered vital for their own security as "[w]ith a viable Japanese economy, it is highly improbable that Soviet efforts at ideological penetration would have any substantial prospect of success."[141] However, much of Northeast Asia was under direct or indirect control of the Soviets, making it impossible for Japan to access these markets for trade. Consequently, the only way to rehabilitate the Japanese economy was for the United States to provide economic aid, even if it meant underwriting substantial trade deficits on an ongoing basis.

2.2 The containment and the American food aids

> Food is the real problem. Empty stomachs mean Bolsheviks. Full stomachs mean no Bolsheviks. […]. We must provide the food […]. We will have to furnish more food than we are now furnishing.

With these words, in 1918, the Secretary of State of the United States, Robert Lansing, expressed the necessity of providing assistance to the populations of Austria and Hungary in alleviating hunger as a means of preventing the proliferation of Communist ideology.[142] After the conclusion of World War I, the United States recognized the valuable tool they possessed in the form of food surpluses, which could be utilized not only to maintain domestic stability, but also as a powerful instrument of foreign policy capable of deterring Communist revolutionary aspirations. Nonetheless, the economic revival of Europe in the 1920s, along with an upsurge in agricultural production on the continent, and the advent of the Great Depression in 1929, brought to a close the American ambitions to feed the world with food produced on their lands. The government intervened

141 Central Intelligence Agency, *Strategic Importance of Japan*, April 1, 1948, p. 8.

142 Robert Lansing, *Robert Lansing Papers: Private Memoranda, 1922*. Originals; 1918 to 1919. Manuscript/Mixed Material. <https://www.loc.gov/item/mss29454002/>

with subsidy measures to prevent a collapse in prices that could have had dire consequences for producers. Additionally, the Smoot-Hawley Tariff Act of 1930 increased tariffs on imported products. In 1933, the Agricultural Adjustment Act introduced measures aimed at incentivizing a reduction in production and subsequently increasing the prices of agricultural products, in an effort to solve what came to be known as "the farm problem."[143]

The outbreak of World War II gave the United States an opportunity to make use of its food surpluses. During the war, the idea that a well-nourished world was crucial for peace and stability gained renewed popularity. On January 6, 1941, in a historic speech to the Congress, President Franklin D. Roosevelt presented the renowned "four freedoms" as a framework for reconstructing the world following the conclusion of the war. The third one, the "freedom from want" referred to the idea that everyone should have access to basic needs such as food, shelter, and clothing, regardless of their economic or social status. Roosevelt believed that achieving freedom from want would not only improve the quality of life for individuals but also contribute to global peace and stability. With the aim of realizing this objective, he called for the United Nations Conference on Food and Agriculture in Hot Springs, Virginia, from May to June 1943, amidst the ongoing global conflict. This assembly ultimately led to the establishment of the Food and Agriculture Organization (F.A.O.) of the United Nations in 1945. However, the scale of the postwar humanitarian and economic crisis was beyond the expectations of policymakers, and millions of people in Europe and around the world faced immediate shortages of food and other basic survival needs. The global disruption of agricultural production caused by the war was so severe that the United States was the only country producing more food after the war than before. In this context, the provision of food aid would have ensured not only the possibility of helping populations struggling with hunger but also of disposing of the agricultural surpluses that had accumulated over the years. But for the United States, the need for postwar relief was becoming entangled in the need for an overarching grand strategy to guide its support for Europe.

"Bare subsistence meant hunger; hunger meant Communism."[144] Echoing Lansing's words, former U.S. President Hoover already in May 1945, shortly after Germany's defeat and still in the midst of the war with Japan, warned President Truman that the hunger prevailing among the populations of the

---

[143] Lewis C. Gray and Oliver E. Baker, *Land Utilization and the Farm Problem*, United States of American, U.S. Department of Agriculture, 1930.

[144] Cited in: Barry Riley, *The Political History of American Food Aid: An Uneasy Benevolence.* New York: Oxford University Press, 2017, p. 109.

countries at war was likely to pose a danger perhaps even worse than Nazism, namely Communism. However, the initial approach of the Allied Powers toward both Germany and Japan was one of retribution and abstention. As discussed in Chapter 1, the prevailing notion was that these nations should shoulder the responsibility for their own economic recovery, including resolving critical food shortages. However, this punitive approach soon proved untenable as humanitarian crises mounted and Cold War realities began to reshape global priorities.

A pivotal moment came with Joseph Stalin's speech in Moscow on February 9, 1946, in which he attributed the origins of World War II to capitalism and signaled the Soviet Union's intent to rearm and challenge the capitalist-imperialist order. This speech served as a stark warning to the United States that, while the Second World War had ended, the seeds of a new ideological conflict were already taking root.[145] Just days later, on February 22, George F. Kennan, an American diplomat, sent from his post at the United States Embassy in Moscow to the Department of State in Washington, a confidential message which would later be referred to as "the Long Telegram." In the telegram, Kennan provided a detailed analysis of the Soviet Union's foreign policy and argued that the Soviet Union was ideologically committed to spreading Communism and would pose a significant threat to the Western powers for the foreseeable future. He then advocated for a policy of containment, which aimed to prevent the spread of Soviet influence without resorting to direct military confrontation. The Long Telegram formed the foundation for the later Truman Doctrine, which sought to halt the expansion of the Soviet Union's sphere of influence by preventing friendly countries from falling to Communism. In this emerging Cold War context, it became essential to rehabilitate the two former adversaries, Germany and Japan, by abandoning the punitive approach that had been upheld until that point and by providing assistance to fully restore their economies. Food aid emerged as a cornerstone in this strategy. The first public presentation of the idea of an American aid program to give food and capital goods to speed European and Japanese recovery was given by Undersecretary Dean Acheson on May 8, 1947, at Delta State Teachers College in Cleveland, Mississippi. He pointed out that Europe and Asia had exhausted themselves in the war and, as of spring 1947, Germany and Japan, the great workshops of prewar Europe and Asia, had not yet begun the essential process of economic rebuilding necessary for the recovery of their respective regions.[146] For this reason, according to Acheson, it was not only necessary for America to provide Europe and Japan with the food they needed, but it also had to commit

---

[145] Ibid., p. 118

[146] Ibid., p. 140.

to importing as much as possible from these regions. This was in order to give them the opportunity to repay the loans they had taken from the United States and international financial organizations for their own reconstruction.[147]

The European Recovery Plan, commonly referred to as the "Marshall Plan," epitomized the manifestation of the new American approach in foreign policy. Ultimately, trough the Marshall Plan, the United States distributed over $13 billion in aid to 16 European countries between 1948 and 1951, helping to create the conditions for postwar economic recovery and political stability. But, more than anything, it aimed to prevent the spread of Communism by promoting economic stability and democracy in Europe.[148]

Japan was another critical element of the evolving American strategy during the early Cold War. As outlined earlier, the precarious conditions of the Japanese population during the early months of the Occupation had incited social unrest, to the point that the fear of a Communist revolution was no longer perceived as a remote possibility by American and Japanese authorities. Meanwhile, on the continent, nationalist forces in China were losing ground to the Communists in 1947, signaling a potential realignment of power in the region. Recognizing the dual threats posed by the international situation and the growing strength of Japanese left-wing parties, U.S. authorities revised their approach to Japan, and implemented measures aimed at improving its economy and society. Among the most significant of these was the Agrarian Reform (*Nōchi kaikaku*), enacted in October 1946. As part of the broader Allied effort to dismantle the concentration of wealth and power held by a few large landowners, this reform sought to create a more equitable distribution of land. It also aimed to revive Japanese national agricultural production to alleviate the problem of food shortage and countering the spread of subversive ideas. As MacArthur said, the land reform "formed a strong barrier against any introductions of Communism in rural Japan."[149] However, it would have taken considerable time to achieve any tangible outcome, and the mounting pressure from civil society made it increasingly urgent to provide foreign food aid to ease the situation.[150]

---

147 Ibid., p. 141.

148 Bryan L. McDonald, *Food Power: The Rise and Fall of the Postwar American Food System*. New York: Oxford University Press., 2017, pp. 36–7.

149 McArthur, *Reminiscences*, p. 314.

150 A few years later, the Japanese Land Reform was regarded as one of the great successes of the American strategy to contain the spread of Communist ideas through agricultural development. See: Psychological Strategy Board, "Inventory of Cold War operations detailed," GALE|CK2349324991, United States: White House, October 11, 1951, p. 4.

Meanwhile, the international stage was undergoing dramatic shifts, with tensions between the United States and the Soviet Union steadily escalating. In the Soviet Union, the efforts to develop the atomic bomb were rapidly advancing, culminating in its first successful test during the summer of 1949. Simultaneously, in China, the Communist forces were prevailing over the nationalists, ultimately leading to the proclamation of the People's Republic of China, in October 1949. In this context, the Central Intelligence Agency (C.I.A.) predicted that a war between the United States and the Soviet Union could erupt within a period of two years. In such a volatile context, it was crucial that Japan, given its industrial potential and strategic significance for the security of the United States, did not fall into the Soviet orbit. Controlling Japan was considered the key to controlling a war economic system in East Asia. However, Japan's value depended on its ability to procure necessary raw materials, including food; otherwise, it would be more of a liability than a resource. Thus, it was the responsibility of the United States to maintain a "political atmosphere favorable to the U.S. through substantial economic assistance."[151]

The evolving international context and the new strategic importance of the Japanese archipelago in the American containment strategy finally led to the initiation of much-desired aid to assist Japan in recovering from a persistent food crisis that had been affecting the nation for years so far. Ultimately, the United States successfully persuaded even its most rigid allies, who had a stronger inclination toward adopting a punitive approach against their former adversary, to implement several food aid initiatives aimed at assisting the Japanese population.

2.3 Food aids to Japan: L.A.R.A., G.A.R.I.O.A., and E.R.O.A.

When the first relief supplies from the Licensed Agencies for Relief in Asia, also known as L.A.R.A., were distributed among the Japanese,

> "everyone rejoiced, as if a beam of light had shone through. The stomachs were pleased. The hands were pleased. The whole bodies were pleased. Light shone in the hearts. It was a sense of gratitude. Great gratitude."[152]

---

[151] C.I.A., *The Strategic Importance of the Far East to the U.S. and U.S.S.R.* Document number: CIA-RDP78-01617A003400130008-2. Publication date: May 4, 1949, p. 14.

[152] "Lala no messeji (Messages for Lara)," URL: <https://www.hcrjapan.org/lara/>, extracts from: Tatara Toshio, *Kyūen no busshi wa taiheiyō o koete: Sengo Nihon to Lala no katsudō* [Relief Supplies Cross the Pacific: Postwar Japan and L.A.R.A.'s Activities], Tokyo: Hoken fukushi kōhō kyōkai, 1999.

L.A.R.A. was an aid organization established in March 1946 by the Committees for Japan and Korea of the American Council of Voluntary Agencies for Work Abroad (A.C.V.A.). Its primary objective was to serve as a coordinating body for various voluntary agencies involved in relief efforts within the occupied countries. A similar organization, known as C.R.A.L.O.G. (Committee of Relief Agencies Licensed to Operate in Germany), had been previously formed by the A.C.V.A., and shipments to Germany commenced in April 1946. The two organizations shared almost the same members and consisted primarily of associations involved in service and reconstruction activities in Germany along with Christian religious groups, predominantly Protestant, involved in relief activities.[153]

In April 1946, S.C.A.P. authorized three representatives from L.A.R.A.—Ernest Bott from Church World Service, Esther B. Rhoads from the American Friends Service Committee, and Father Harold J. Felsecker from the War Relief Service—to travel to Japan and work as intermediaries between the S.C.A.P. and the overseas sending agencies.[154] They soon reached an agreement permitting the importation of relief supplies to be distributed through Japanese government welfare channels under the supervision of the S.C.A.P.[155] On August 30, 1946, the S.C.A.P. issued a memorandum—"Receipt and Distribution of Relief Supplies from Licensed Agencies for Relief in Asia" (SCAPIN 1169)—to the Japanese government with the aim of guaranteeing a correct receipt and distribution of L.A.R.A. relief supplies to those in need. Shortly thereafter, on September 20, the Japanese government, in response to the S.C.A.P.'s requests, presented the "General Plan Regarding Receipt and Distribution of L.A.R.A. Relief Supplies." These documents outlined the legal framework for L.A.R.A.'s activities in Japan and provided guidelines for how relief supplies should be obtained and distributed. Essentially, they established the rules and procedures that governed L.A.R.A.'s humanitarian work in the country.[156] On November 30, 1946, the first L.A.R.A. aid ship arrived at the port of Yokohama, carrying essential supplies including milk, rice flour, butter, jam, and clothing, totaling nearly 500 metric

153 The members included among others: American Friends Service Committee; Church World Service, Inc.; Luthern World Relief, Inc.; Salvation Army; War Relief Service; Young Men's Christian Association; Young Women's Christian Associations. See: *Ibidem*, "Lala kyūen katsudō no hottan wa nani ka [What was the origin of L.A.R.A.'s relief activities?]."

154 Ibidem.

155 Ibidem.

156 Ministry of Health and Welfare (Kōseishō), *Lara kinenshi* [L.A.R.A. memorial book], Tokyo: Kōseishō, 1996, pp. 56–57.

tons of relief supplies.[157] After one year, L.A.R.A. had distributed 5,000,000 pounds of foodstuffs, valued at $1,512,000. According to Ernest Bott, American aid had been "a paramount factor in restoring self-respect to the Japanese people and in heightening their admiration for the democratic process."[158] From the initial delivery until the cessation of L.A.R.A.'s operations in 1952, approximately 33 million pounds of aid were delivered to Japan. Of this total, approximately 75 percent comprised food supplies.[159]

While relief supplies from private charitable organizations played a crucial role in alleviating the difficulties of the neediest segments of the population, the financial support provided by the government of the United States was instrumental in Japan's long-term recovery from food shortages. At the beginning of the 1947 U.S. fiscal year (July 1946–June 1947), the United States established the Government and Relief in Occupied Areas (G.A.R.I.O.A.) program, which aimed to provide support to Japan and Germany in order to prevent famine, disease, and social unrest within these countries. Through this program, the United States were able to provide essential resources such as food, fertilizers, and medical supplies. Following the G.A.R.I.O.A. program, the United States also established the Economic Rehabilitation in Occupied Areas (E.R.O.A.) program during the 1949 U.S. fiscal year (July 1948–June 1949). Aside from ensuring the provision of vital necessities, the E.R.O.A. program served also the purpose of furnishing industrial raw materials and machinery essential for the economies of the two defeated nations. The United States continued to provide assistance to Japan through these two programs until June 1951.[160]

A closer examination of G.A.R.I.O.A. and E.R.O.A. reveals two distinct operational phases. The first, prior to April 1, 1949, focused primarily on stabilizing Japan's domestic economy and promoting exports.[161] As we have already pointed out, the primary objective was to ensure economic stability by utilizing the aid goods from the United States to maintain a low and stable cost of living, thereby

---

157 Yomiuri shinbun, "Ajia kyūzai kikan Lara. Dai ikkai no okurimono. Sensaisha to gakudō kyūshoku ni [Asia relief organization L.A.R.A. First gifts to war victims and schoolchildren's meals]," 30 November 1946.

158 Anonymous, "$1,512,000 is spent on food for Japan," *The New York Times*, 13 September 1947.

159 Ministry of Health and Welfare, *Lara kinenshi*, pp. 73–4.

160 Takagi, Shinji, *From recipient to donor: Japan's official aid flows, 1945 to 1990 and beyond*, Essay in International Finance, Princeton: Princeton University, 1995, p. 6.

161 Committee on Appropriation—United States Senate, "Hearings—First Session on H.R. 4830 A Bill making Appropriations for Foreign Aid for the Fiscal Year Ending June 30, 1950, and for other purposes," Washington D.C.: U.S. Government Print Office, 1949, p. 807.

preventing disease outbreaks and social unrest. During this initial phase, it was possible to leverage the absence of a fixed exchange rate and the existence of multiple exchange rates. It was then possible to bring into Japan G.A.R.I.O.A. and E.R.O.A. goods and sell them at controlled prices that corresponded to the lower price levels established for similar domestically produced items. For example, the official prices of locally produced food, which set the foundation for pricing food imports, were deliberately maintained at relatively low levels compared to the general price level and industrial production costs. This approach, known as "pricing in" imports at low yen prices, aimed to maintain price and wage stability by not reflecting the higher postwar dollar prices of Japanese imports. Given that food prices exerted significant influence on wages and the prices of other commodities in an economy heavily reliant on subsistence, this strategy served the purpose of "holding the line" on price and wage adjustments.[162] The yen proceeds generated from the sales of G.A.R.I.O.A. and E.R.O.A. supplies were deposited by the United States into a foreign trade fund, which predominantly served as a financial resource for the acquisition of domestically manufactured products in Japan. This system facilitated the gradual reestablishment of Japanese foreign trade and enabled the Japanese government to acquire funds for the purchase of additional imported food and raw materials.[163] Due to the existence of divergent yen-to-dollar exchange rates, where imports were sold at lower ratios and goods for export were acquired at higher ratios, the yen proceeds obtained from the sales of G.A.R.I.O.A. and E.R.O.A. goods simultaneously fulfilled the dual objectives of ensuring price stability in the domestic market and maintaining competitive export prices in world trade.[164]

The economic reforms introduced by Joseph Dodge in 1949, as part of his role as a financial advisor to the Japanese government, significantly reshaped the distribution and utilization of aid provided through these programs.[165] Specifically, two key reforms had a direct impact on the G.A.R.I.O.A. and E.R.O.A. programs. Firstly, a fixed exchange rate between the dollar and the yen was established at a rate of 360 yen per dollar. Secondly, a special import account was established

---

162 Ibidem, p. 807–8.

163 Ibidem.

164 Ibidem.

165 In December 1948, the U.S. government sent Joseph Dodge, the President of Detroit Bank, to Japan as a financial advisor to help end Japan's persistent inflation and stabilize the economy. In March 1949, he presented the key points of his proposals, based on the principles of balanced budgets, orthodox finance, stabilized prices, and a fixed exchange rate at 360 yen per U.S. dollar. See: Shimamoto, Mayako, et al., *Historical dictionary of Japanese foreign policy*, Lahnam, Maryland: Rowman & Littlefield, 2015, pp. 62–3.

to which aid counterpart funds were credited. On April 1, 1949, in response to Dodge's policies, S.C.A.P. sent a memorandum to the Japanese government, requesting the establishment of a special account at the Bank of Japan, which would serve as the United States aid counterpart fund for the stabilization of Japanese economy.[166] In this account, the Japanese government would transfer sums equivalent to the monthly yen value of G.A.R.I.O.A.-E.R.O.A. imports at the rate of 360 yen per dollar. As a result, on April 25, the Japanese government approved the "United States-Japan Aid Counterpart Fund Special Accounting Law (*Beikoku tai-nichi e enjo mikaerishikin tokubetsu kaikeihō*)," which established the "counterpart fund" ("mikaerishikin" in Japanese). This law, which consisted of 16 articles, established a U.S.-Japan aid counterpart fund to account for yen repayments. The term "counterpart fund" referred to the unique situation where the aid goods, originally denominated in U.S. dollars, were being repaid in yen. The Japanese government was required to obtain S.C.A.P.'s approval for each planned expenditure from the counterpart fund. S.C.A.P. would have authorized releases from the counterpart fund only for specific reasons, which were limited to: (a) the repayment of public debt; (b) capital improvements in designated government enterprises, such as railways and communications; and (c) capital improvements in key private industries. In general, disbursements from the fund were required to contribute to economic recovery and were contingent upon the Japanese government or the specific industry demonstrating satisfactory progress in enhancing administrative and production efficiency.[167] This mechanism allowed the United States to exercise control over the items, quantity, and even the use of market sale proceeds of aid goods.

The provision of official U.S. aid to Japan through G.A.R.I.O.A. and E.R.O.A. programs continued until the end of June 1951. Due to the predominant administration of the assistance by S.C.A.P., acquiring precise and separate data regarding the exact amount of aid received by Japan during this period is difficult. According to some estimations, Japan received approximately $2 billion in total aid under these two programs.[168] It is important to note that, despite the formal name of "aid," these funds were not intended as unconditional donations. The United States, in fact, expected that Japan would, albeit under advantageous conditions,

---

[166] S.C.A.P., "Memorandum for Japanese Government," in Committee on Appropriations, "Hearings," pp. 808–9.

[167] Ibidem.

[168] Takagi, *From recipient to donor*, p. 7; Takita, Kenji, "Kokusai shakai to Amerika no senryō tainichi keizai enjo—G.A.R.I.O.A., E.R.O.A. enjo wo chūshin toshite [The International Community and Postwar U.S. Economic Assistance to Japan—A Focus on G.A.R.I.O.A. and E.R.O.A.]," in *Hōgaku shinpō*, March 2015, p. 327.

repay the amounts received through the two programs. The negotiations for the repayment of the funds started shortly after Japan regained its independence in April 1952.

During the meeting between Ikeda Hayato, the then Chairman of the Liberal Party's Political Affairs Research Council, who visited the United States as Prime Minister Yoshida Shigeru's private envoy, and Walter S. Robertson, the United States Assistant Secretary of State for Far Eastern Affairs, the matter of G.A.R.I.O.A.-E.R.O.A. was brought to the forefront. Within the discussions, it was acknowledged that the Japanese economy still exhibited a considerable degree of instability, and the fulfillment of G.A.R.I.O.A.-E.R.O.A. claims would impose a lasting burden on Japan's economy for an extended period of time. The negotiations for the repayment of G.A.R.I.O.A.-E.R.O.A. funds lasted approximately ten years thereafter. Finally, in June 1961, Ikeda Hayato, who had assumed the role of Prime Minister by then, and President Kennedy of the United States issued a Memorandum of Understanding, which eventually paved the way for the repayment agreement.[169] The agreement, known as "Agreement between the United States and Japan for the Settlement of Postwar Economic Assistance to Japan" or in Japanese "Nihon ni taisuru sengo no keizai enjo no shori ni kansuru Nihonkoku to Amerika gasshūkoku to no aida kyōtei," was officially signed in Tokyo on January 9, 1962. According to the terms of the agreement, Japan committed to repay $490 million to the United States over a span of fifteen-year period (Article 1), with the provision to defer or suspend payments in the event of adverse economic circumstances (Article 2).[170] Additionally, it was also decided in a attached protocol that the funds repaid by Japan would have primarily used by the United States to finance economic assistance provided to developing nations and a specific allocation of $25 million was designated for fostering educational and cultural exchange initiatives between Japan and the United States.[171]

---

169 "Settlement of the United States Claim for Postwar Economic Assistance to Japan (G.A.R.I.O.A.)," in *Foreign Relations of the United States 1961–1963*, Volume XII, Northeast Asia, Document 353, Washington, March 22, 1962. URL: <https://history.state.gov/historicaldocuments/frus1961-63v22/d353>

170 The full text of the agreement is available at: Ministry of Foreign Affairs (Gaimushō), *Diplomatic Blues book, 1962*, URL: <https://www.mofa.go.jp/mofaj/gaiko/bluebook/1962/s37-shiryou-2.htm#1>

171 Ibidem.

2.4 The Mutual Security Act

During the negotiations for the peace treaty that would restore Japan's sovereignty and mark the end Allied Occupation, significant disagreements emerged over Japan's future economic trajectory. Once again it was the United Kingdom, in particular, that voiced reservations regarding Japan's future economic trajectory, primarily fearing Japan's resurgence as a commercial competitor in East Asia. These apprehensions led to a British proposal aimed at exerting control over Japanese economy. Central to this proposal was the suggestion to deprive Japan of its substantial gold reserves, valued at approximately $200 million at the time, and incorporate the country into the sterling area. This arrangement would have effectively placed Japan under British economic supervision.[172] In his capacity as the negotiator for the preliminary discussions on the Japanese treaty, John Foster Dulles strongly opposed the British proposal. He underscored that, during the Occupation, the United States had refrained from appropriating Japan's gold reserves, even though it could have done so to offset the costs of Occupation. Instead, American policy had prioritized Japan's economic recovery and self-sufficiency, with the aim of relieving the country from being a burden on other nations.[173] According to Dulles, the rebuilding of Japanese industry and the revival of its export capacity were indispensable. Without a functioning economy, Japan would be unable to procure essential commodities such as food and raw materials.[174]

During the Occupation, the United States had taken the responsibility of providing Japan with the necessary food. However, upon achieving independence, it was essential not to deprive Japan of its resources through the payment of monetary reparations, otherwise "her ordinary commercial credit would vanish, the incentive of her people would be destroyed and they would sink into a misery of body and spirit which would make them an easy prey to exploitation," and eventually the "old menace would appear in aggravated form."[175] Ultimately, the peace treaty negotiations resulted in the formulation of Article 14, which allowed Japan to meet its war reparations obligations by providing machinery and other industrial equipment rather than making monetary payments. This arrangement aimed to strike a balance between acknowledging the need for reparations and ensuring Japan's post-Occupation economic revival.

---

172 Committee on Foreign Affairs, *U.S. policy in the Far East, Part I, Vol XVIII*, Washington: U.S. Government Printing Office, 1980, pp. 11–12.

173 Ibidem, p. 12.

174 John F. Dulles, *A Peace Treaty in the Making*, Japanese Peace Conference, San Francisco, September 4–8, 1951, Washington D.C.: U.S. Government Printing Office, p. 11.

175 Ibidem, p. 19.

The intensification of the Cold War rivalry between the two blocs prompted Washington to acknowledge the impracticality of addressing multiple strategic theaters simultaneously. As a result, the United States prioritized efforts to bolster the military capacities of its allies, including Japan, to distribute the security burden more equitably. In this context, during the negotiations for Japan's peace treaty, the U.S. government introduced the Mutual Security Act (M.S.A.), formally enacted as Public Law 165 and signed by President Truman on October 10, 1951. The M.S.A. was a cornerstone of American strategy, designed to facilitate the reconstruction and enhancement of military capabilities among Allied nations. The M.S.A. represented a significant financial commitment, with $7.5 billion allocated for military, technical, and foreign aid initiatives. Politically, the Act aimed to strengthen the cohesion of the American-led alliance system. Its provisions included the supply of military weaponry and equipment, training programs for Allied officers within the United States, and the establishment of mechanisms to ensure closer coordination of military strategies among member states.[176]

In Japan, discussions regarding the Japanese-American Mutual Defense Assistance Agreement commenced in September 1952, shortly after the establishment of the National Safety Force (*Hōantai*). However, the negotiations faced significant obstacles, as thoroughly documented by John Welfield.[177] On one side, Japanese officials were markedly hesitant to expand their military capabilities or allocate increased resources to defense expenditure. They argued instead for the redirection of M.S.A. funds toward stimulating economic growth, reflecting a prioritization of postwar economic recovery over militarization. Conversely, American representatives remained steadfast in their insistence that the M.S.A. funds be used exclusively for security-related purposes. They viewed the enhancement of Japan's defense infrastructure as a critical element of regional stability and a necessary contribution to the broader containment strategy against Soviet influence.

During an initial meeting held in Mexico City, Ikeda Hayato, representing Japan, engaged with key American officials, including John W. Snyder, the United States Secretary for the Treasury, and Joseph Dodge, the State Department Special Adviser. The American delegation pressed for a substantial expansion of Japan's National Safety Force, proposing an increase from the existing 110,000 personnel

176 Martin Folly, *Historical Dictionary of U.S. Diplomacy During the Cold War*, Lanham, Maryland: Rowman & Littlefield Publishers, 2014, p. 275.

177 John Welfield, *An Empire in Eclipse: Japan in the Postwar American Alliance System: A Study in the Interaction of Domestic Politics and Foreign Policy*, London: Bloomsbury Publishing, 2013, pp. 97–108.

to 180,000 personnel by the end of 1952. They further outlined an ambitious long-term objective to establish a formidable 325,000-strong army by 1954–1955. To facilitate this expansion, the American representatives expressed their willingness to seek congressional approval for a budget allocation of $300 million, aimed at equipping the enlarged Japanese force with the necessary armaments and equipment. However, the Yoshida government responded with hesitation, expressing difficulty in surpassing the existing strength of 110,000 personnel for Japanese forces.[178]

For a time, Prime Minister Yoshida was able to resist American pressures. However, mounting domestic and international factors began to erode this position. Domestically, Japan faced an economic downturn following the cessation of United States Korean War special procurements, which had previously provided a significant economic stimulus. This prompted business leaders and other stakeholders to advocate for the acceptance of M.S.A. aid. Supporters of the aid argued that, if carefully negotiated, it could channel substantial resources toward Japan's economic reconstruction, fostering technological advancement and enhancing industrial competitiveness. Conversely, rejecting the aid risked straining Japanese-American relations and undermining Japan's trade ties with U.S. allies in Southeast Asia and the Pacific.[179]

To alleviate Japanese concerns, American officials reassured Yoshida's government that Japan would retain a degree of flexibility in negotiating the scope and type of assistance received. They further emphasized that the aid could have a positive impact on Japan's economy without imposing excessive military obligations.[180] These assurances gradually softened Japanese resistance, as policymakers began to perceive M.S.A. aid as beneficial both economically and militarily, without necessitating significant new military commitments. Following additional encouragement from the United States, Japan expressed its willingness to accept M.S.A. assistance. However, Okazaki, on June 16, clarified to the Diet that the focus of the agreement was on the economic aspects rather than the military ones. Finally, on June 30, the Cabinet made the decision to apply for aid, and two weeks later, on July 15, the first round of negotiations started.[181]

The negotiations that began in July 1953 extended over a protracted and contentious period, lasting more than eight months and revealing deep-seated differences between Japan and the United States. At the heart of the dispute

---

178 Ibidem, p. 98.
179 Ibidem, p. 99.
180 Ibidem.
181 Ibidem.

lay diverging perspectives on the relationship between M.S.A. assistance and economic stability. Japan sought to prioritize economic stability within the agreement, aiming to include provisions that addressed its economic concerns, particularly the purchase of surplus agricultural products by the United States. In contrast, the American delegation resisted this approach, maintaining that M.S.A. assistance should remain focused on military objectives. In particular, Japan, with a strong determination to include economic assistance all costs, focused on the provision in the M.S.A. that pertained to the U.S. purchase of surplus agricultural products.[182]

From August 14 to September 30, 1953, a total of seven negotiation sessions took place. During this time, Japan presented a draft that included various economic requests, reflecting its determination to secure economic aid under the M.S.A. framework. However, American negotiators rejected these requests, arguing that the program was explicitly limited to military aid and not intended to address economic development.[183] By late September, the negotiations persisted with minimal advancement on foundational matters, but among in Washington, the idea of offering some concessions to the Japanese started to gain ground. A breakthrough came during the Ikeda-Robertson negotiations in October 1953, when the two sides reached a pivotal agreement concerning the provisioning of agricultural commodities. Under the accord, $50 million worth of agricultural products was allocated to Japan under Section 550 of the M.S.A. It was also established that the generated proceeds from the sale of these agricultural commodities, denominated in local currency, were to be strategically channeled toward augmenting the defense production capabilities and industrial prospects of Japan.[184] While the sum may have been modest, the accomplishment of obtaining defense support aid under the guise of economic aid marked a significant achievement for Japan.

In the following months, negotiations centered on specific logistical and economic aspects related to the trade of agricultural products. Specifically, the main issues revolved around the selling price, the type of wheat, and the transportation methods. It has been estimated that the debate around the economic aspects of the M.S.A. took up three-quarters of the entire negotiation time, prolonging the

182 Ishii, Susumu, "M.S.A. kyōtei to Nihon. Sengo kei Keizai shisutemu no keisei (2) [The M.S.A. Agreement and Japan: Formation of the Postwar Economic System]," in *Gakushūin daigaku Keizai ronshū*, vol. 40, n. 4, 204, p. 297.

183 Ibidem.

184 "Editorial Note," in *Foreign Relations of the United States, 1952–1954, China and Japan*, Vol. XIV, part 2, document n. 713, URL: <https://history.state.gov/historicaldocuments/frus1952-54v14p2/d713>

M.S.A. talks beyond expectation.[185] Finally, the "U.S. and Japan Mutual Defense Assistance Agreement" was signed on March 8, 1954, in Tokyo between the American Ambassador in Japan, John Moore Allison, and the Japanese Foreign Minister, Okazaki Katsuo. On the same day, four additional bilateral agreements were signed in Tokyo, among which the "Agreement between Japan and the United States of America regarding the Purchase of Agricultural Commodities."[186] This agreement facilitated the acquisition of agricultural commodities, primarily wheat, from the United States, amounting to a total value of $50 million. As it was agreed during the Ikeda-Robertson talks, these agricultural products would be sold to Japan using local currency, and the resulting funds would be used to support military expansion efforts. Out of the $50 million generated from food sales within Japan, $40 million was allocated for the purchase of military material for United States military aid programs, while the remaining $10 million was to be available as counterpart funds for defense support to Japan.

The signing of the treaty provoked intense debate in Japan, particularly concerning the potential risks that could have arisen. In particular, two aspects were deemed particularly problematic. The first one was the relation between the M.S.A. agreement and the Japanese Constitution. An editorial in the Yomiuri Shinbun highlighted unresolved constitutional doubts, despite prior Diet discussions. The M.S.A. agreement outlined Japan's military responsibilities in a more explicit way than the San Francisco Treaty, potentially indicating a deeper military involvement with the United States. This raised fears of scenarios outside the constitutional framework, such as overseas troop deployments.[187] The second point concerned the economic aid, which did not meet Japanese expectations. The only notable positive aspect was the allocation of ten million dollars as counterpart funds for the development of Japan's defense industry, although its significance was deemed relatively modest.[188] However, the issue regarding the utilization of these ten million dollars also became the subject of prolonged and divergent negotiations between the Americans and the Japanese. According to the agreement, the conditions governing the use of these counterpart funds were to be agreed upon by the Japanese and United States governments. As of June 1954, the two parties had not yet reached an agreement on how to use these funds. On one hand, the United States wanted to use them for financing the

185 Ishii, "M.S.A. kyōtei," 2004, p. 303.

186 The other three agreements were: the "Agreement (with exchange of notes) regarding the guarantee of investments," the "Agreement on Economic Arrangements," and an agreement on "Arrangements for the Return of Equipment under the M.D.A.A."

187 "Shasetsu," Yomiuri Shinbun, 9 March 1954.

188 Ibidem.

defense industry, while on the other, Japan continued to propose projects related to general economic development, such as highway construction.[189]

In July 1954, the Minister of Agriculture Hori Shigeru had a conversation with the U.S. Agricultural Attachés in Tokyo, Wolf I. Ladejinsky, the "chief architect" of the land reform, and Joseph C. Dodson, during which he expressed strongly criticized the tying of $10 million grant to defense purpose only. He alleged that this had an unfortunate effect upon United States Japanese relations, and expressed the hope that the error would not be repeated in the future.[190] He even went so far as to say that the allocation of these funds for defense was the United States' greatest postwar mistake, second only to the immediate release of Japanese Communists at the outset of the Occupation.[191] Hori underscored the concerns of Japanese farmers regarding the adverse effects of excess agricultural products and stressed the necessity for tangible actions against the Communists ("the common enemy") who were disseminating anti-American propaganda in Japan. The measure he had in mind was to use the yen derived from the sale of American surplus agricultural commodities toward the increase of domestic agricultural production.[192] At that time, the Japanese "understood that aid, trade, and military policy were inseparable" but at a certain point they thought that "this linkage had become too intimate, too military, too cumbersome."[193]

The negotiations around the counterpart funds continued for several more months, and in November 1954, Ambassador Allison, in a letter addressed to Noel Hemmendinger, the Director of the Office of Northeast Asian Affairs at the Department of State, suggested considering the idea of withdrawing any purely military-related proposals, otherwise they would consistently face strong opposition from the Japanese.[194] He recommended that the funds be used to finance

---

189 "Foreign Service Dispatch," June 29, 1954, *Records of the U.S. Department of State Relating to the Internal Affairs of Japan* (RDOS/IAJ), 794.5, MSP/6-2954, n. 1727, Reel 19.

190 "Memorandum of conversation with the Minister of Agriculture and Forestry, Mr. Shigeru Hori, and other high ranking Ministry officials concerning the use of yen proceeds of surplus American Agricultural commodities sales to Japan for agricultural development in Japan (with Wolf I. Ladejinsky and Joseph C. Dodson)," July 14, 1954, *RDOS/IAJ*, 894.20/7-1554, Reel 35.

191 Ibidem.

192 Ibidem.

193 John W. Dower, *Empire and Aftermath: Yoshida Shigeru and the Japanese Experience, 1878–1954*, Cambridge, Massachusetts: Council on East Asian Studies, Harvard University, 1979, p. 480.

194 "Telegram 1054. American Embassy in Tokyo to Noel Hemmendinger, Director, Office of Northeast Asian Affairs, Department of State," November 10, 1954, *RDOS/IAJ*, 794-5 MSP/11-1054, Reel 19.

those activities involved in the production of parts, components or major items of equipment not required in sufficient quantities to justify commercial production or in the development of designs and methods of manufacture that will permit Japan's industry to go into munitions production promptly in the event of war.[195]
Eventually, a compromise was reached, making it possible to finance several projects, including those involving Fuji Communication Apparatus Mfg. Co. for the production of condensers and carbon resistances, essential to electronics industry, as well as Osaka Kingoku Kōgyō Co. for the fluorine resin process, important to defense industry.[196] Other projects involved the revitalization of the Japanese aircraft industry. One particularly significant example is the Japan Jet Engine Company, a semi-official enterprise established in July 1953 and later reorganized in August 1954 through the allocation of part of the M.S.A. wheat grant.[197] Other companies that received substantial financial support from the M.S.A. wheat grant included the Asahi-Okuma Industry, engaged in ammunitions production, and the Tokyo Precision Machine Co.[198]

The M.S.A. wheat agreement ultimately proved mutually beneficial for both the United States and Japan, serving distinct yet complementary objectives. For the United States, the agreement provided an effective mechanism to offload its surplus agricultural commodities while simultaneously strengthening its strategic alliance with Japan as a bulwark against the spread of Communism in Asia. For Japan, the arrangement offered a crucial opportunity to import affordable foodstuffs, thereby conserving its limited dollar reserves. These savings could then be redirected toward acquiring essential resources to fuel the country's nascent economic recovery and industrial development. In this way, the agreement not only alleviated immediate economic pressures but also contributed to Japan's broader postwar reconstruction and long-term growth.

2.5 The Public Law 480

As discussed, the M.S.A. aid program was predominantly designed for military purposes, making it an ill-suited mechanism for addressing the problem of the growing agricultural surpluses in the United States. Concurrently, while agricultural production had rebounded robustly in many war-damaged nations, broader indicators

---

195 "Telegram 1142. From Tokyo to the Secretary of State," October 29, 1954, *RDOS/IAJ*, 794.5 MSP/11-954 Reel 19.

196 "Telegram 1495. From Tokyo to Secretary of State," December 24, 1954, *RDOS/IAJ*, 794.5 MSP/12-2454.

197 *Flight, The Aircraft Engineer*, vol. 67, p. 413.

198 *American Machinist*, n. 99, p. 157.

of economic growth remained stagnant. A notable impediment to the economic expansion within most European nations and Japan resided in the insufficiency of foreign currency required for the acquisition of essential capital commodities. One of the criticisms directed at the Japanese was their excessive concentration on importing goods aimed at enhancing the standard of living rather than procuring materials to channel into export markets.[199] Assistance provided through earlier programs such as E.R.O.A. and G.A.R.I.O.A. had indeed materially contributed to the economic recovery. However, by 1953, Japan faced a sharp economic downturn exacerbated by a poor rice harvest, driving domestic prices to unsustainable levels. This left Japan struggling to afford even basic imports.[200] It was in this context that the Eisenhower administration enacted the Agricultural Trade and Development Act, better known as P.L.480, on July 10, 1954.

The Act consisted of three titles. Title I authorized sales of American farm products to friendly nations in exchange for foreign currencies. These funds could be used for several purposes, including developing new markets for United States agricultural goods, purchasing strategic materials, acquiring military equipment, fostering economic development, and supporting international educational exchange. Title II provided for emergency food aid during crises such as famines, while Title III governed internal and foreign donations and barter transactions involving strategic materials.[201]

The enactment of P.L.480 reflected the convergence of three main factors. First, there was a prevailing willingness among public opinion and policymakers to continue extending aid to both Europe and less developed nations. Simultaneously, the costs of storing surplus agricultural products had become difficult to sustain, necessitating innovative disposal strategies. Moreover, U.S. trading partners faced a scarcity of foreign exchange, limiting their ability to purchase American goods.[202] P.L.480 addressed these issues by enabling recipient countries to pay for U.S. agricultural products in their local currencies, which the United States reinvested domestically within the recipient country under mutual agreements. This approach was beneficial to both sides. The United States could dispose of

---

199 "An economic program for Japan," August 1954, *RDOS/IAJ*, FOA 894.00/8554, Reel 26.

200 "Foreign Service Dispatch. Japan's Economic Position and Balance of Payments," August 4, 1954, *RDOS/IAJ*, 894.00/8-454, Reel 26.

201 Public Law 480. Full text available at: <https://www.gpo.gov/fdsys/pkg/STATUTE-68/pdf/STATUTE-68-Pg454-2.pdf>

202 Amalia Vellianitis-Fidas and Eileen Marsar Manfredi, *P.L.480 Concessional sales—History, Procedures, Negotiating and Implementing Agreements*, Foreign Agricultural Economic Report No. 142, Foreign Demand and Competition Division, Economic Research Service, U.S. Department of Agriculture, December 1977, p. 3.

its agricultural surpluses, while the recipient countries benefited by saving their own foreign currencies, which could be used for buying other necessary items from abroad.[203]

Beyond domestic and trade objectives, P.L.480 also contributed to achieving foreign policy objectives. The legislation explicitly stated that the foreign currency received by the United States would have been used "to promote collective strength, and to foster in other ways the foreign policy of the United States."[204] In particular, P.L.480 was conceived as an additional tool within the containment strategy. In fact, the law, in its original form, restricted aid to "friendly nations" only, which were defined as "any country other than (1) the U.S.S.R., or (2) any nation or area dominated or controlled by the foreign government or foreign organization controlling the world Communist movement."[205] Thus, it was stated that the President had the authority to assist friendly nations to be independent of trade with the U.S.S.R. or nations dominated or controlled by the U.S.S.R. for food, raw materials, and markets, and to assure that agricultural commodities sold or transferred under the P.L.480 did not result in increased availability of those or like commodities to unfriendly nations.

In the specific context of Japan, P.L.480 was strategically significant. Within the U.S. Department of State, concerns were mounting that Japan's reliance on food imports could lead it to align more closely with the Soviet Union. As Robert R. Mullen, Assistant Secretary for Economic Affairs at the Department of State, wrote in response to President Eisenhower's call for major government departments to find a solution to Japan's economic recovery while adhering to the four truths[206] of American diplomacy:

---

203 Ibidem.

204 Public Law 480. URL: <https://www.gpo.gov/fdsys/pkg/STATUTE-68/pdf/STATUTE-68-Pg454-2.pdf>

205 Ibidem.

206 The "four truths" illustrated by Eisenhower were: "1. The United States cannot carry all the nations of the world on its shoulders; we should stop give away programs. 2. We cannot allow nations with whom we are allied to trade with the Reds, providing them with strategic materials with which to destroy us. 3. We can't defend the whole world and its freedom, especially any part that doesn't really want freedom. 4. We can't trade with countries whose living standards are so far below ours that they hurt our industries, mines, shops or farms." Based on these "truths," he asked to the government agencies: "Just apply those four points to Japan … if we will not give her money, if we will not trade with her, if we will not allow her to trade with the Reds, if we will not try to defend in any way the Southeast Asia area where she has a partial trade opportunity, what is to happen to Japan?." From: "Letter from Robert R. Mullen to Mr. Samuel C. Waugh, Assistant Secretary for Economic Affairs, Department of State," August 11, 1954, *RDOS/IAJ*, 894.00/8-1754, Reel 26.

> Japan is on the brink of a decision affecting the world's power balance. The jargon of the economists can be used to describe Japan's crisis, but *if you were to reduce the problem to one word, that word would be food [emphasis added]*. And it is equally and starkly apparent to all who study the problem that Japan will eat and thus survive either under U.S. or Soviet auspices.[207]

Mullen was aware that the Japanese had initiated unofficial discussions with Moscow aimed at resuming their trade relations. He was also aware that the Russians would be willing to sell their grain to Japan, potentially leading to a heightened level of political and even military engagement.[208] For the United States, it was therefore imperative to prevent this rapprochement and to avoid "losing Japan," thereby averting the transformation of the Pacific into a "Communist lake."

The concern that Japan's food crisis could profoundly influence its alignment within the free world was echoed by the American Ambassador to Tokyo, John M. Allison. In a letter addressed to the Vice President Richard Nixon, Allison emphasized that the stability of Japan's economy was of fundamental importance to the United States in order to prevent Japan "from slipping behind the 'bamboo curtain,' of keeping it enrolled in the cause of the free world."[209] Specifically, he recommended, among other things, the utilization of America's surplus agricultural products to extend crucial assistance to Japan amidst its currency predicaments. Given Japan's limited food resources and shortage of foreign currency, he also suggested that the United States offered "generous terms of sale including a substantial portion of the yen proceeds as a grant-in-aid," in order "to assist Japan and at the same time dispose of its agricultural surpluses."[210] In this context, the utilization of P.L.480 appeared to be the solution capable of reconciling both the domestic needs of Japan and the United States, as well as diplomatic and international considerations.

By September 1954, the Japanese government had begun drafting a P.L.480 Japanese proposal, with the objective of attaining a comprehensive agreement during Prime Minister Yoshida Shigeru's visit to Washington in November. The proposal was articulated on October 14, 1954, by the Economic Planning Agency[211]

---

[207] Ibidem.

[208] Ibidem.

[209] "The Ambassador in Japan (Allison) to Vice President Nixon," July 8, 1954, in *Foreign Relations of the United States, 1952–1954, China and Japan*, vol. 14, part 2, document 775, URL: <https://history.state.gov/historicaldocuments/frus1952-54v14p2/d775>

[210] Ibidem.

[211] "Yojō nōsanbutsu kaiire narabi ni kaiire kawari en shikin no shiyō ni tsuite [Purchase of Surplus Agricultural Products and Replacement Yen Funds]," J.A.C.A.R., Ref. B23010150500, E'.2.3.1.4_007, p. 0160.

and transmitted by the Japanese government to the United States Government agencies on October 24, under the title "Basic Concepts of the Government of Japan Regarding the Acquisition and Disposal of U.S. Surplus Agricultural Commodities."[212] The proposal included the acquisition of several commodities, including wheat, barley, non-fat milk powder for animal feed, soybeans, raw cotton, leaf tobacco, and corn, amounting to a cumulative sum of $133.39 million. In this proposal Japan expressed the will to utilize the yen proceeds generated from the purchase of surplus agricultural products from the United States "to attain the required economic stability and defensive strength, to render her invulnerable to internal subsession and external aggression." To achieve this goal, it was deemed crucial to promptly acquire a substantial amount of investment funds. Therefore, the proposal requested that all the yen funds obtained from the sale of American agricultural products in Japan be entirely allocated to projects aimed at enhancing the nation's strength and resilience. These projected encompassed a range of strategic investments including: the establishment of a robust export market for Japanese manufacturers through investments in the Southeast Asian countries; the support of some agricultural development projects within Japan, such as the Aichi irrigation project and the land development initiatives in Hokkaidō; the improvement of defense industry and the construction of the defense road network; the establishment of a Production Promotion Center with the aim of disseminating new and more efficient techniques and methods to Japanese businesses and industries; the support of the emigration of Japanese to other less densely populated areas of the world. Finally, a part of these funds would be used for the financing of a school lunch program designed to improve the nutritional standards of Japanese children.[213]

During Prime Minister Yoshida Shigeru's visit to Washington in November 1954, the two countries reached a general agreement regarding the allocation of $100 million in surplus agricultural commodities for Japan under P.L.480, with $85 million falling under Title I and an equivalent of $15 million under Title II. In his memoirs, Yoshida Shigeru recalls how both parties had agreed on the mutual benefits of a deal, given that the United States needed to dispose of its agricultural surpluses while Japan required food. Once certain points were clarified, negotiations proceeded smoothly, resulting in a mutually satisfying agreement for all involved

---

212 "Basic Concepts of the Government of Japan concerning Japanese Acquisition and Disposal of U.S. Surplus Agricultural Commodities," Document n. 10225, J.A.C.A.R., B23010151200, E'.2.3.1.4_008, p. 0185.

213 Ibidem, p. 0186.

parties.[214] The joint statement by Yoshida and President Eisenhower reflected this spirit of cooperation and satisfaction with the reached agreement, which was poised to enhance Japan's economy and bolster its defense capabilities.[215] However, outside official communications, there was a pronounced discontent among certain members of the Japanese government regarding the reached agreement. Aichi Kiichi, Minister of International Trade and Industry who was conducting the negotiations with the Americans, described the agreement as "far from satisfactory," and in a written response stated that this "would seem to suggest that approximately 55 percent of the Japanese currency generated by sales of surplus commodities should be used for the purposes which could be of no benefit to Japan and which, in fact, might prove to be detrimental to our general objective of strengthening our country so that it might be an effective partner in the free world alliance."[216] They demanded that a minimum of 80 percent of the P.L.480 yen proceeds, which amounted to approximately $70 million, be allocated to projects in line with the proposals put forth by the Japanese government itself.[217] After Miyazawa Kiichi, a member of the House of Councilors and a member of the Aichi delegation, expressed "strong dissatisfaction" too, the United States raised the amount of uncommitted proceeds to $50 million on the day before Yoshida's departure.[218] From this point on, the negotiations became more difficult, and there were doubts on the U.S. side about Japan's commitment. In March, there were discussions about possibly stopping the negotiations because the economic benefits were limited, and there could be political consequences. However, the negotiations continued and on May 31, 1955, the "Agreement on Agricultural Commodities between Japan and the United States of America" (*Nōsanbutsu ni kan suru Nihonkoku to Amerika gasshūkoku to no aida no kyōtei* in the Japanese version) was officially signed in Tokyo.[219] The final text confirmed the American proposal: $85 million granted under Title I of P.L.480, plus an additional $15 million under Title II, the latter to be used to expand school lunch programs in Japan (article III).

Despite the dissatisfaction with an agreement that fell below expectations, the Japanese government, now guided by Hatoyama Ichirō, immediately initiated

---

214 Yoshida Shigeru, *The Yoshida Memoirs. The Story of Japan in crisis*, London: William Heinemann Ltd., 1961, p. 122–4.

215 "Text of Joint Japanese-American Statement," November 10, 1954, J.A.C.A.R., Ref. B2301015140, E'.2.3.1.4_008, p. 0418

216 Quoted in Dower, *Empire and Aftermath*, p. 484.

217 Ibidem.

218 Ibidem.

219 The full text of the agreement is available at: www.mofa.go.jp/mofaj/gaiko/treaty/pdfs/A-S38(3)-260.pdf

negotiations for the conclusion of a second agreement for the purchase of American agricultural surpluses. Minister of Agriculture Kōno Ichirō led the team negotiating for Japan. By September 1955, they managed to create the first draft of the agreement. Although both sides tried not to make the same mistakes as in the first negotiations and things went more smoothly, there were some points that caused a lot of discussion, particularly concerning rice imports, which were ultimately excluded from the agreement.[220] The final agreement, which was signed in Tokyo on February 10, 1956, outlined that Japan committed to purchasing U.S. agricultural commodities worth $65.8 million under Title I of P.L.480. Simultaneously, the United States agreed to provide Japan with non-fat dried milk and wheat valued at $15.0 million under Title II for their school lunch program.[221] After the implementation of the second agreement, discussions for a third agreement began but failed twice due to changing circumstances on both sides, including Japan's improved domestic production and shifts in America's policies.[222]

One of the consequences resulting from the acquisition of American agricultural surpluses was a change in the dietary habits of the Japanese population. The three treaties (1954, 1955, and 1956), as previously mentioned, dealt with the supplies and financing of the Japanese school lunch program. In particular, the imported ingredients were used to meet the goals set by a 1951 directive of 600 kilocalories and 25 grams of protein, to be provided by a roll, a glass of milk, and a warm dish.[223] The adoption of these products led to an increased consumption of dairy and wheat-based products among Japanese students. Rice, on the other hand, largely disappeared from school lunches and only made a return starting in 1976.[224] If the new school lunch menus aimed to introduce Japanese children to new flavors from a young age, it was also necessary for mothers to learn how

---

220 For the detail of negotiations see: Itō Atsushi, "Dai 2ji PL480 kyōtei o meguru Nichibei kōshō (1955–56 nen). Nichibei nikokukan kankei o koete [U.S.-Japan negotiations under the Agricultural Trade Development and Assistance Act (Pl480), 1955–1956: beyond bilateral relations]," in *Shakai Keizaishi-gaku*, vol. 88, n. 1, May 2022, pp. 33–55.

221 The full text of the agreement is available at: www.mofa.go.jp/mofaj/gaiko/treaty/htmls/A-S38(3)-261.html

222 For the detail of negotiations see: Itō Atsushi, "Dai 3ji dai 4ji PL480 kyōtei o meguru Nichibei kōshō (1956–57 nen). Seiritsu shinakatta kyōtei no kōshō kara no shōsha [U.S.-Japan negotiations under the Agricultural Trade Development and Assistance Act (Pl480), 1956–1957: from the viewpoint of rejected agreements]," in *Shakai Keizaishi-gaku*, vol. 89, n. 1, May 2023, pp. 27–52.

223 Katarzyna J. Cwiertka, *Modern Japanese Cuisine: Food, Power and National Identity*, London: Reaktion, 2006. p. 163.

224 Stephanie Assmann, "The remaking of a national cuisine: The food education campaign in Japan," in James Farrer (ed.) *The globalization of Asian cuisines: Transnational networks and culinary contact zones*, New York: Palgrave Macmillan U.S., p. 178.

to cook new recipes with these new ingredients. For this purpose, the kitchen cars program was implemented. Kitchen cars were vans provided by the Oregon Wheat Growers League and equipped with kitchens inside, which were used to introduce Japanese housewives to new ingredients and their preparation methods. The first demonstration took place in Tokyo, in Hibiya Park, in October 1954, and from then until the end of the 1960s, approximately 12 kitchen cars traveled extensively throughout Japan, offering demonstrations of new recipes.[225]

Although some contemporary voices may cast doubt on the influence of American food aids on Japanese eating habits,[226] it is clear that Japan experienced a profound shift in its food culture and nutritional levels starting in the 1950s. Beginning in these years, something significant occurred: for the first time since the postwar era, Japanese daily calorie consumption exceeded 2000 kcal and the calories supplied (*kyōkyū karorī*) far surpassed those actually consumed (*sesshu karorī*), highlighting the shift from the food-deficient nation Japan once was just a few years earlier.[227] Notable changes occurred also in food consumption patterns. Specifically, aided by economic growth and an improved standard of living, there was a decline in the consumption of rice, a traditional staple in Japan, while there was a notable increase in the consumption of meat, fats, and dairy products.[228]

2.6 Conclusion

When General MacArthur granted his permission to import food and the first aids arrived in Japan, a wave of jubilation spread throughout the country. During the summer matsuri celebrations in July and August of 1946, it was possible to read everywhere signs of gratitude with the inscriptions "Amerika arigatou (Thank you, America)" or in English "In appreciation of General MacArthur's sincere Aid for Japan's Food Crisis."[229] The composers Norita Kōzō and Soma Mitsuo even wrote the "Gratitude Song for Imported Food and Supplies [*Yunyū shokuryō*

[225] See: Fujiwara, *Kyūshoku no rekishi* p. 159–162; Nathan Hopson, "Ingrained habits: the 'kitchen cars' and the transformation of postwar Japanese diet and identity," in *Food, Culture & Society*, vol. 23, n. 5, 2020, pp. 589–607.

[226] Itō Atsushi, "Japanese Market Development Programs under P.L.480 in the late 1950s: Focus on Wheat and Rice," in *Agricultural History*, 2021, vol. 95, n. 2, pp. 245–275.

[227] <https://honkawa2.sakura.ne.jp/0203.html>

[228] Kusakari Hitoshi, "Sengo no shokuryō shōhi to kakei no yakuwari [Postwar Food Consumption and the Role of Household Budgets]," in *Japanese journal of research on household economics*, n. 83, p. 13.

[229] Hiratsuka Masao, *Shashin de wakaru jiten no senryōshi: 1945nen 8gatsu—1952nen 5gatsu* [English title: History of occupied Japan: Aug. 1945–May 1952], Kyoto: Php Editors, 2019, p. 112–13.

busshi kansha no uta]" which was first presented during the "Gratitude Event for Imported Food and Supplies [*Yunyū shokuryō busshi kansha no kai*]" in Aomori City, organized by the Tōō Nippō Newspaper Company, in September 1946.[230]

The local administrations also made sure to convey their profound appreciation directly to General MacArthur for his pivotal role in alleviating the food crisis. This was the message addressed to the General by the Osaka administration on August 15, 1946:

> The unprecedented food crisis and the dire state of poverty that our nation faces have brought us to the brink of starvation. In this challenging time, the Supreme Commander for the Allied Powers has taken every possible measure to provide relief, including the release of food imports from the United States. We, the 130,000 citizens, are deeply grateful and moved by this opportunity. We will embrace the spirit of humanitarianism, based on friendship and mutual assistance, to share our limited resources and address the deficiencies. We aim to repay the generosity of the Supreme Commander by overcoming this food crisis. Therefore, the Municipal Council, in particular, resolves to express our heartfelt gratitude to the Supreme Commander for the Allied Powers.[231]

As discussed in the previous chapter, the United States' decision to accede to Japanese requests for food aid was primarily driven by a humanitarian imperative, reflecting President Roosevelt's ideal of achieving "freedom from want." Despite critiques—particularly those accusing the United States of employing food aid as a "wheat strategy" to westernize Japanese diets and erode traditional food practices—, this program undeniably played a transformative role in improving the nutritional health of the Japanese population. As Nakamura Teiji observed, "Japan changed the U.S. food policy into a nutrition policy that made people healthier," marking a pivotal transition away from the widespread malnutrition that characterized the immediate postwar years.[232]

Beyond humanitarian goals, the formalization of food aid through programs such as G.A.R.I.O.A., E.R.O.A., and later bilateral agreements like the M.S.A. and P.L.480 underscored a complex interplay of economic and security interests. On one hand, the United States was able to develop new markets for its commercial surpluses to avert an economic crisis that would have global repercussions, while simultaneously ensuring economic and social stability in countries they deemed vital for national security. The result was that Japan became "the single largest

---

230 Fujiwara, Tatsushi, *Kyūshoku no rekishi* [The history of school lunch], Tokyo: Iwanami Shoten, 2018, p. 101.

231 Taiheiyō sensō kenkyū-kai, *G.H.Q. no mita Nippon* [Japan seen by the G.H.Q.], Sekaibunkasha, 2007, p. 45.

232 Nakamura, Teiji. *Japan Nutrition*, Singapore: Springer Nature, 2021, p. 45.

market for U.S. agricultural commodities."[233] On the other hand, Japan, no longer able to rely on imports from its former colonies or domestic production, had the opportunity to meet the needs of its hungry population and, at the same time, obtain investments to revive its industrial sector. Negotiations between the two nations reveal the depth of Japanese agency in these agreements, challenging narratives that frame them solely as instances of an American "wheat strategy." While the United States emphasized rearmament within the broader framework of Cold War geopolitics, Japan strategically advocated for allocating proceeds from these agreements toward economic development projects. This duality of priorities underscores the extensive dialogue and mutual accommodation that shaped these arrangements.

Finally, from a security and international perspective, it was mutually acknowledged that enhancing the nutritional and economic well-being of Japan's civilian population constituted a pivotal element in suppressing social upheaval and halting the spread of revolutionary ideologies that posed a threat to Japan's alignment with the bloc of free nations.

[233] Theodore W. Galdi, *Use of U.S. Food Resources for Diplomatic Purposes: An Examination of the Issues*, Washington, D.C.: United States Government Printing Office, 1977, p. 17.

CHAPTER 3

Pressing Japan: *Gaiatsu* and the "Soybean Embargo"

3.1 Introduction

The conclusion of the Allied Occupation of Japan exacerbated the challenges posed by the nation's scarcity of natural resources. This issue was compounded by the American trade embargo on the People's Republic of China and North Korea, cutting Japan off from key suppliers of raw materials. While the inflow of American surplus goods partially mitigated the challenges posed by a domestic production system struggling to meet the population's needs, it became imperative for Japan to re-engage with the international market to secure essential resources. This reintegration ignited a profound and multifaceted debate, both within Japan and beyond its borders. Domestically, the discourse revolved around the perceived risks associated with further opening up the country to international trade, particularly concerning agricultural products. Some argued that such a move could have detrimental effects on the already vulnerable domestic agricultural sector. Conversely, there were proponents of increased liberalization who saw it as not only a means to secure vital imports but also as a strategy to bolster Japanese manufacturing industries' access to foreign markets. Externally, the debate was similarly divided. Some nations feared that a more open Japan might inundate global markets with competitively cheap products, potentially harming their own economies. On the other side, others advocated for Japan's increased engagement with the international community. In particular, in Japan, skepticism regarding the liberalization of domestic market was primarily fueled by fears of a "Black Ships" scenario, wherein the United States and other Western countries would exert control over Japan's domestic market.[234]

[234] Takahashi Kazuhiro, "Ikeda seiken-ki ni okeru bōeki jiyūka to nashonarizumu [Trade Liberalization and Nationalism during the Ikeda Administration]," in *Kokusai seiji*, n. 170, 2012, p. 47.

The lengthy process of Japan's accession to the General Agreement on Tariffs and Trade (G.A.T.T.) is representative of this complex interplay of diverse viewpoints and interests, both within Japan and among its trading partners. From Washington's perspective, it was fundamental to "prevent Japan from becoming dependent upon China and other Communist-dominated areas for essential food and raw material supplies through stimulation of Japan's trade with other free nations."[235] When Japan applied for membership to the G.A.T.T. in 1952, it received strong support from the United States, which had already made attempts in 1948 and 1949 to bring Japan into the organization.[236] On the other side, the main obstacle to Japan's adhesion to G.A.T.T. was the opposition of the United Kingdom, who was strongly concerned about Japan's trade practices such as copying British designs, engaging in price-cutting strategies, and having lower wages and labor standards.[237] Internally, Japan's political landscape was marked by divisions between the ruling conservatives and the opposition socialists. The conservative camp, led by figures like Yoshida Shigeru and Hatoyama Ichirō, saw G.A.T.T. membership as a means to reintegrate Japan into the international community and promote foreign trade recovery. On the other hand, the socialists were critical of American capitalism and advocated expanding trade with the Communist bloc, particularly the People's Republic of China and Southeast Asia.[238] On September 10, 1955, these divisions were finally overcome, and Japan was admitted to the G.A.T.T., marking the end of the recovery period and the beginning of an era of growth.[239] However, this was just the first step toward greater openness to international trade. In the second half of the 1950s, Japan's balance of payments displayed such a substantial surplus that the international community exerted pressure on the Japanese government to eliminate its trade barriers.[240] As we will see in this chapter, in the early 1960s, in response to these external pressures (*gaiatsu*), Japan adopted a liberalization program that led the

---

235 National Security Council 125/2, in *Foreign Relations of the United States, 1952–1954, China and Japan*, vol. XIV, part 2, Document n. 588, Washington, August 7, 1952, URL: <https://history.state.gov/historicaldocuments/frus1952-54v14p2/d588>

236 Aaron Forsberg, "The Politics of G.A.T.T. Expansion: Japanese Accession and the Domestic Political Content in Japan and the United States, 1948–1955," in *Business and Economic History*, Fall 1998, vol. 27, n. 1, p. 186.

237 Aaron Forsberg, "Eisenhower and Japanese Economic Recovery: The Politics of Integration with the Western Trading Bloc, 1952—1955," in *The Journal of American-East Asian Relations*, Spring 1995, vol. 5, n. 1, p. 67.

238 Forsberg, 1998, pp. 188–90.

239 Forsberg, 1995, p. 75.

240 Aurelia George Mulgan, "The role of foreign pressure (gaiatsu) in Japan's agricultural trade liberalization," in *The Pacific Review*, vol. 10 n. 2, 1997, p. 174.

country to increase its liberalization rate from 41 percent in 1960 to 93 percent in 1965.[241] The liberalizations also affected the agrifood sector, leading to a significant increase in imports. During the period from 1960 to 1967, there was a substantial annual rise of 18.1 percent in the importation of food products, a notable increase if compared to the 1.5 percent annual rate recorded in the 1950s.[242]

In the 1960s, the global economy's favorable conditions led to a positive outlook on increased international trade openness. However, the economic crises of the early 1970s undermined confidence in international trade, prompting many states to adopt robust protectionist measures. In the case of agrifood trade, we will analyze how the so-called "soybean embargo" imposed by the Nixon administration in 1973 significantly affected Japan, as soy was a crucial ingredient in the nation's diet and Japan was heavily dependent on imports from the United States. The consequence of the embargo was Japan's pursuit of a new strategy aimed at ensuring a steady supply of food. This involved investing in developing countries and increasing domestic production to reduce dependency on the United States, a reliance that, as we have seen, had developed since the early postwar period.

Focusing on these two main events, the liberalizations of the 1960s and the food crisis of the 1970s, we will see how during this period the trade relationship between Japan and the United States experienced profound tensions. These tensions arose from American pressure to open the Japanese market, only to later "betray"—to use the term employed by American historian and former U.S. ambassador to Japan Edwin O. Reischauer to describe how Japan's government felt toward Washington—their ally during the "soybean embargo."[243] We will explore how this "betrayal" was the catalyst for Japan's change in strategy. For the first time since the postwar period, Japan decided to pursue a path of reduced dependence on the United States and seek new sources of supply through a reconceptualization of the concept of national security.

3.2 *Gaiatsu* or the new black ships

Starting from the mid-1950s, just a decade after the end of World War II, Japan embarked on a long period of high economic growth, commonly referred to as the "period of high economic growth (*kōdo keizai seichō jidai*)." For more than

---

[241] Alfred K. Ho, *Japan's Trade Liberalization in the 1960s*, New York: International Arts and Science Press, 1973, p. 16.

[242] Ibidem, p. 34.

[243] Edwin O. Reischauer and Marius B. Jansen, *The Japanese Today. Change and Continuity*, Cambridge: Harvard University Press, 1999 (original printing 1977), pp. 368–9.

15 years, until the beginning of the 1970s, the Japanese economy kept growing at a remarkable average rate of 8.7 percent per year, occasionally even exceeding 12 percent.[244] During this period of high economic growth, in the late 1950s, Japan experienced a substantial surplus in its international balance of payments. However, this surplus attracted strong criticism from the international community, which urged Japan to enhance the accessibility of its domestic market. In August 1959, the American ambassador to Japan, Douglas MacArthur II, nephew of General MacArthur, expressed serious concerns about Japanese-American economic and trade relations during a conversation with Fujiyama Aiichirō, the Japanese Minister of Foreign Affairs. The ambassador stressed that Japan's exports to the United States had increased by 51 percent in the last six months, while imports from the United States had decreased by 2.7 percent during the same period, leading to the United States experiencing an adverse trade balance with Japan for the first time in the postwar era. He emphasized the urgent need for Japan to address these issues by promoting liberalization and facilitating freer trade with the United States.[245] In response to these pressures, on September 16, 1959, during the Economic Cabinet Members Consultative Conference (*Keizai kakuryō kondakai*), the Japanese government decided to transition nine out of the ten items[246]—with the exception of iron ore—covered by the Fund Allocation System (F.A. system) to the Automatic Approval system (A.A.) and liberalize them by the end of 1960.[247]

---

[244] Itami Hiroyuki, "Chapter 4. Economic Growth and Japanese Management," in *Seven Chapters on Japanese Modernization*, J.I.C.A.-Open University of Japan, BS231, August 2019, p. 1.

[245] "Telegram from the Embassy in Japan to the Department of State," in *Foreign Relations of the United States, 1958–1960, Japan, Korea*, vol. XVIII, Document n. 94, Tokyo, August 27, 1959, URL: <https://history.state.gov/historicaldocuments/frus1958-60v18/d94>

[246] The items included scrap iron, iron ore, abaca fibers, gypsum, beef tallow, lard, soybeans, copper alloy scrap, meranti wood, and cattle hides. See: "Jū hinmoku no yunyū jiyūka [The liberalization of import of ten items]," *Yomiuri Shinbun*, 16 September 1959.

[247] The F.A. system (*gaika shikin wariate sei*) was the primary method of import control in Japan, where quotas were established for importing specific goods from particular currency areas. It applied to essential items, like food and raw materials, primarily from the dollar area. Importers received allocation certificates to obtain foreign exchange. The A.A. system (*jidō shōnin sei*) had an overall limit on imports from a currency area but did not impose specific limits on individual commodities. Licenses were granted as long as the budgetary quota wasn't fully used, providing flexibility for various imports. In 1953, the F.A. system was applied to about 65 percent of all imports while the A.A. system was applied to 34 percent. See: Takagi, Shinji, "Japan's Restrictive System of Trade and Payments Operation, Effectiveness, and Liberalization, 1950–1964," I.M.F. Working Paper 1997, p. 11.

The 15th session of the G.A.T.T., hosted in Tokyo from October 26 to November 20, 1959, served as the primary forum for discussions on Japan's international trade. The central concern for Japan revolved around Article 35 of the General Agreement on Tariffs and Trade, which enabled specific member countries to discriminate against Japanese imports. Despite Japan's G.A.T.T. membership since 1955, 14 countries, including the United Kingdom and France, invoked Article 35, leading to discriminatory practices against Japan and urging Tokyo to eliminate dumping measures and open its market. The Japanese representative, Abe Isao, emphasized that Japan was facing significant challenges such as economic weaknesses, population and employment issues, and an unfavorable trade environment, and he argued that Japan needed to approach liberalization cautiously, highlighting Japan's progress in reducing discriminatory import restrictions and pledging to remove all discrimination on imported items in the near future, while also mentioning plans to increase global quotas and transfer more items to the Automatic Approval system, with the goal of providing more opportunities for the import of manufactured goods.[248]

The issue of liberalization became the central point of conversations between Under Secretary Douglas Dillon, who was in Tokyo for the G.A.T.T. session, and Satō Eisaku, who was then the Minister of Finance in the government of Kishi Nobusuke (1957–1960). During the meeting, Satō provided assurance that the Japanese government was fully committed to progressing a program of liberalization, particularly in relation to scrap iron and specific agricultural products, like soybeans. While acknowledging the Japanese efforts, Dillon expressed some reservations regarding the timeline for the implementation of these liberalization measures and requested the Japanese to make as many efforts as possible to expedite the implementation of this program. It is noteworthy, however, that the primary motivation behind the American sense of urgency was not solely to attain immediate economic advantages but rather to secure a positive diplomatic outcome ahead of the presidential elections in the autumn of 1960.[249]

At the conclusion of the G.A.T.T. 15th session, Japan committed to significant steps toward a substantial liberalization. It was announced that, out of the ten items (namely lauan, scrap copper alloy, abaca fiber, and gypsum), which were subject to import restrictions from dollar areas under the foreign currency allocation

---

[248] G.A.T.T., "Summary Records of the Tenth Meeting," Tokyo, 5 November 1959, S.r.15/10. URL: <https://docs.wto.org/dol2fe/Pages/SS/directdoc.aspx?filename=Q:/GG/SR/15-10.PDF&Open=True>

[249] "Memorandum of Conversation," in *Foreign Relations of the United States, 1958–1960, Japan, Korea*, vol. XVIII, Document 102, Tokyo, October 28, 1959, URL: <https://history.state.gov/historicaldocuments/frus1958-60v18/d102>

system, four would see their import restrictions lifted, commencing in January 1960. The remaining six items would have their import restrictions removed as expeditiously as possible in the subsequent year. Additionally, it was announced that 177 other items would be reclassified as Automatic Approval items or be subjected to alternative liberalization measures, effective in January.[250]

Under such pressure from the international community, on January 5, 1960, the Cabinet decided to establish the Ministerial Meeting for the Promotion of Trade and Foreign Exchange Liberalization (*Bōeki kawase jiyūka sokushin kakuryō kaigi*) to elaborate the basic policies on trade and exchange liberalization to be implemented by May of the same year.[251] In the same month, during his visit to Washington, Prime Minister Kishi reiterated to the Secretary of State, Christian A. Herter, the firm intention of resolving the balance of payments issue with the United States. He stressed that Japan would continue to make "as much haste as possible to remove such discriminatory measures."[252] With the aim of further encouraging the Japanese government to expeditiously implement a liberalization program, the U.S. government dispatched a trade mission to Japan in the spring of 1960. The mission, comprising two officials from the Department of Commerce and four American entrepreneurs, was active in Japan from March 9 to April 28. During this period, they traveled across the country to showcase American products. The mission had a dual objective: firstly, to increase general interest in Japan regarding American products and, secondly, to prompt the Japanese government to relax import restrictions as soon as possible.[253]

Despite American efforts and pressures, Japan's liberalization plan took more time than initially projected, and it was not completed until June 24, one month past the established deadline. The "Outline Plan of the Trade and Exchange Liberalization [*Bōeki kawase jiyūka keikaku daimō*]" defined the objective of substantially increasing the liberalization rate over a span of three years. However, it also stipulated that liberalization would be executed incrementally, taking into account the specific circumstances of each industry. The plan categorized commodities into four distinct groups for the purposes of liberalization: 1)

---

[250] "MITI says four import items to be freed in January," *The Japan Times*, 12 November 1959.

[251] Arisawa, Hiromi and Inaba Hidezō, *Shiryō. Sengo nijūnen-shi. Vol. 2, Keizai* [Documents. History of the twenty years after the war. Volume 2. Economy], Tokyo: Nippon Hyōronsha, 1966, p. 369.

[252] "Memorandum of conversation," in *Foreign Relations of the United States, 1958–1960, Japan, Korea*, Document 139, Washington, January 19, 1960. URL: <https://history.state.gov/historicaldocuments/frus1958-60v18/d139>

[253] "Hearings before the United States Senate Committee on Appropriations," H.R. 10753, Washington: United States Government Printing Office, 1960, pp. 528–534.

commodities to be immediately liberalized; 2) commodities to be liberalized within three years; 3) commodities that could not be liberalized immediately in the near future (more than three years); and 4) commodities that could not be liberalized in the foreseeable future.[254]

The plan was indeed ambitious for Japan; nonetheless, it drew criticism from the United States that found the pace of liberalization outlined in the plan to be excessively slow. During a meeting in Washington on September 12, 1960, shortly after the Ikeda administration took office in Japan, Undersecretary L. Merchant urged Japan to accelerate its trade liberalization efforts, otherwise, there was a risk of the United States imposing restrictive measures on imports from Japan, a demand increasingly pressed by American industries.[255] Japanese Foreign Minister Kōsaka Zantarō emphasized the government's strong commitment to trade liberalization under the Ikeda administration and promised that in the following three years, the number of liberalized goods would increase from 40 to 80 percent. However, due to the unique characteristics of the Japanese economy, marked by a significant number of small businesses with fewer than 30 employees and a substantial portion of the population engaged in the agricultural sector (approximately 38 percent), it would have been difficult to accelerate the process of opening up the market.[256] However, despite these difficulties, Japan upheld its commitments and promptly initiated the first measures aimed at achieving the outcomes outlined in the plan. In September 1960, the Special Exchange Fund Allocation System was abolished. By April 1961, the system of linking raw material imports to exports was also eliminated. On July 1, 1961, any form of discrimination against the dollar area was abolished, with the removal of restrictions on dollar imports of soybeans and refined lard. In terms of import liberalization, more than 700 items were added to the list of goods under the A.A. system from October 1960 to July 1961, bringing the number of items on the A.A. list to 1997, which constituted 65 percent of all imports.[257] Collectively, these measures raised the liberalization index from 44 percent in October 1960 to 65 percent by July 1961.[258] However, international pressures for a more accelerated liberalization process did not relent, leading Japan to formulate a new plan in September

---

254 Arisawa and Inaba, *Shiryō*, pp. 370–2.

255 "Memorandum of conversation," in *Foreign Relations of the United States, 1958–1960, Japan, Korea*, Document 199, Washington, September 12, 1960. URL: <https://history.state.gov/historicaldocuments/frus1958-60v18/d199>

256 Ibidem.

257 Takagi, "Japan's Restrictive System of Trade," p. 24.

258 Takagi, Shinji, *Conquering the Fear of Freedom: Japanese Exchange Rate Policy since 1945*, Oxford, Oxford University Press, 2015, p. 67.

1961, the "Plan for the Enforcement of the Liberalization of Trade and Foreign Exchange" (*Bōeki kawase jiyūka sokushin keikaku*). The new plan has advanced the process of liberalization, setting even more ambitious objectives. While the 1960 plan aimed for an 80 percent liberalization rate to be achieved within three years, the new plan raised the bar, targeting a 90 percent rate to be reached by October 1962.[259] Thus, from October 1961 to April 1962, Japan embarked on a significant effort by opening up more than a thousand items, resulting in a liberalization rate of 73 percent. By October 1962, they further expanded this endeavor by liberalizing an additional 230 items, thereby elevating the liberalization rate to 88 percent. This fell just 2 percent short of the target set in September 1961 but exceeded the June 1960 goal by 8 percent. Subsequent measures undertaken in the following years brought the liberalization process close to completion, with the rate reaching 92 percent by the beginning of 1964.[260]

Alongside domestic market liberalization efforts, Prime Minister Ikeda, the man that spent "spent half of his life to the internationalization and liberalization of Japanese economy,"[261] embarked on a diplomatic mission to Europe in the autumn of 1962, seeking to redress the economic restrictions imposed on Japan due to the application of Article 35 of the G.A.T.T. Ikeda's mission concluded with great success. Not only did he establish a trade and commerce agreement with the United Kingdom, initiating negotiations for a similar pact with France, which ultimately culminated in 1963, but he also achieved the cessation of long-standing discriminatory trade practices against Japan, which had persisted for nearly two decades in the postwar era.[262] By 1963, Japan had successfully achieved full-fledged G.A.T.T. membership as an Article 11 country, and the following year, it attained Article 8 status within the I.M.F. Finally, in 1965, Japan was admitted as a full member of the Organization for Economic Cooperation and Development (O.E.C.D.), thereby leaving behind its former status as a developing nation and firmly establishing itself as a developed country.

---

259 "Bōeki kawase jiyūka sokushin keikaku [Plan for the Enforcement of the Liberalization of Trade and Foreign Exchange]," Cabinet Prime Minister's Office, 26 September 1961, URL: <https://www.digital.archives.go.jp/img.pdf/1366486>

260 Takagi, *Conquering the Fear*, p. 67.

261 Miyazawa Kiichi, *Sengo seiji no shōgen* [Testimony of postwar politics], Tokyo: Yomiuri shinbunsha, 1997, p. 129.

262 Oliviero Frattolillo, *Reassessing Japan's Cold War. Ikeda Hayato's Foreign Politics and Proactivism During the 1960s*, London, Routledge, 2019, pp. 52–64.

3.3 The liberalization of soybeans

The issue of agrifood imports was a central and, perhaps, the most important aspect of discussions about liberalizations in Japan. While it was widely acknowledged that Japan needed to increase its import of agrifood products due to limitations in increasing domestic production, it is also true that, despite economic growth, the Japanese economy was still highly reliant on the agricultural sector, which accounted for around 38 percent of the labor force in Japan in 1960.[263] There were fears that major liberalization would put more pressure on a sector which was unable to compete in world markets without government support. On the other hand, for the United States, Japan was their most important market for agricultural products. As stated in a report from the U.S. Senate titled "Japan. The American Farmers' Best Export Market":

> Most American farmers are aware of the importance of foreign markets, but not many realize that their best overseas customer is Japan. They may know that almost half of the U.S. wheat, cotton and soybean crops is exported every year. But few are aware that Japan buys more American cotton, soybeans, hides and skins, and tallow than any other country in the world.[264]

According to the report, in the period 1952–1961 the United States sold more farm products to Japan to any other countries. In 1961, agricultural exports to Japan accounted for more than 11 percent of total U.S. farm exports, with a value of $554 million. More specifically, in 1961 the United States supplied 95 percent of Japan's soybean imports and 30 percent of wheat imports. Additionally, in 1961, the United Stated exported 55.5 million pounds of non-fat dry milk and 27.6 million pounds of raisins to Japan, corresponding to about 22 percent and 23.1 percent of total U.S. dry milk and raisins exports.[265] The report anticipated that Japan would not only continue to be the primary market for the United States in the coming decade but would also undergo further expansion. Economic growth, increasing incomes, and shifts in dietary preferences were all factors that foreshadowed a 61 percent increase in food product imports by 1970.[266] For this reason, it was essential for Japan to maintain its commitment to open its market,

---

263 "Memorandum of conversation," in *Foreign Relations of the United States, 1958–1960, Japan, Korea*, Document 199, Washington, September 12, 1960. URL: <https://history.state.gov/historicaldocuments/frus1958-60v18/d199>

264 U.S. Senate, *Trade Expansion Act of 1962: Hearings Before the Committee on Finance*, United States Senate, Eighty Seventh Congress, Second Session on H.R. 11970, Washington D.C.: United States Government Printing Office, July 1962, pp. 670.

265 Ibidem, pp. 671–677.

266 Ibidem.

and, at the same time, for the United States to increase its imports from Japan to ensure that Japan always had an ample reserve of dollars. Moreover, for the United States, it had become essential to emphasize their export program in response to an unfavorable balance of payments, which resulted from the investments of the Mutual Security Program.[267]

The main issue in the early 1960s concerned the liberalization of soybean imports. Soy has always been a fundamental element of the Japanese diet, as well represented by the numerous dishes derived from it, such as tofu, soy sauce, or miso, which constitute some of the main elements of Japanese cuisine. Furthermore, since the late 1800s, thanks to its characteristics, soy found numerous new uses, ranging from the production of animal feed to the manufacturing of industrial oils, margarines, plastics, and paints. Before the Second World War, Japan had managed to gain control of Manchuria in northeastern China, which was the world's primary region for soybean production at the time. In this way, Japan was able to ensure a steady supply of soybeans.[268] However, the end of the war, as previously discussed in the initial chapter, marked the termination of imports from colonized or controlled regions. Indeed, in 1945, Japan had imported approximately 800,000 tons of soybeans, but in 1946, this figure drastically diminished to a mere 1,582 tons, exclusively supplied by the United States.[269] In the following years, Japan increased its imports from the United States, which remained the primary supplier, while simultaneously attempting to revive domestic production. In 1945, the national soybean cultivation area covered approximately 257,000 hectares, with production reaching 170,000 tons. By the 1950s, production managed to attain an annual average of 450,000 tons over an area of about 400,000 hectares.[270] In 1960, the year prior to the liberalization of soybean imports, the total consumption in Japan was around one and a half million tons, with domestic production accounting for 28 percent of this figure (or 81 percent if we consider only soybeans used for human consumption).[271]

From their perspective, the United States, which had in the meantime become the world's leading soybean producers, surpassing China, saw in Japan

---

267 "Memorandum of conversation," in *Foreign Relations of the United States, 1958–1960, Japan, Korea*, Document 200, Washington, September 16, 1960. URL: <https://history.state.gov/historicaldocuments/frus1958-60v18/d200>

268 Felice Farina, *La via della soia*, Sommacampagna, Cierre Edizioni, 2023, pp.13–29.

269 William Brandemuhl, *Soybean Utilization in Japan*, Volume I, University of Kyoto, Kyoto, 1965, p. 49.

270 Data retrieved from: <https://www.maff.go.jp/j/seisan/ryutu/daizu/d_data/index.html>

271 Inoue Masayuki, "Daizu jiyūka 28 nen wo kataru [Reflecting on 28 Years of Soybean Liberalization]," in *Aff: Agriculture, forestry, fisheries*, vol. 19, n. 12, 1988, p. 55.

a tremendous opportunity for their exports. In order to solidify the trade relationship between the two countries, the Japan American Soybean Institute was established in 1956, bringing together Japanese producers of tofu, miso, and soy sauce on one side and American exporters of soybeans on the other. Another significant event that led to the increase in imports from the United States was the "Nagasaki Flag Incident"[272] in 1958 which led to the suspension of all trade between Japan and the People's Republic of China, from which Japan continued to import a significant amount of soybeans. In this way, the United States went from supplying approximately 75 percent of Japanese soybean imports to 95 percent, while Japan became the largest buyer of American soybeans, accounting for over one-third of all American soybean exports.[273]

The expectations of a significant increase in soybean consumption in the years to come necessitated a greater opening of the Japanese market through the removal of barriers to soybean imports. During his visit to Washington in September 1960, Japanese Foreign Minister Kōsaka, during a talk with the Secretary of Commerce Frederick Mueller, expressed the Japanese government's intention to increase soybean imports from the United States and implement a specific liberalization program for this purpose.[274] However, the implementation of the program did not proceed as fast as Washington expected, and, in October 1960, the Secretary of Agriculture, Ezra T. Benson, was sent to Japan to discuss with officials from the M.A.F.F., in order to address the delay in the liberalization of Japan's soybeans import and push for its prompt realization.[275] The then Vice-Minister for Agriculture, Watanabe Gorō, explained that it was Japan's full intention to implement the liberalization program as soon as possible but, at the same time, they had to ensure that domestic producers would be safeguarded against any shock *(dageki)*.[276] At that time, the Japanese government

[272] The Nagasaki Flag Incident occurred in May 1958 when two Japanese youths pulled down a Chinese flag at a stamp show in a Nagasaki department show. The incident, together with the visit of Prime Minister Kishi Nobusuke in Taiwan, led the Chinese government to suspend all trade relations with Japan.

[273] Nakamura Hiroshi, "The Japanese Soybean Market," in *Illinois Agricultural Economics*, vol. 1, no. 2, 1961, pp. 9–11.

[274] "Memorandum of conversation," in *Foreign Relations of the United States, 1958–1960, Japan, Korea*, Document 200, Washington, September 16, 1960. URL: <https://history.state.gov/historicaldocuments/frus1958-60v18/d200>

[275] "Bei nōmu chōkan 13 nichi rainichi [U.S. Secretary of Agriculture to visit Japan on the 13th]," *Yomiuri Shinbun*, 5 November 1960.

[276] "Nōsanbutsu no kaitsuke—bei nōmuchōkan, zōka o yōsei [Procurement of agricultural products—U.S. Secretary of Agriculture requests an increase]," *Yomiuri Shinbun*, 14 November 1960.

had to mediate between the interests of farmers, who opposed greater market liberalization, and those of soybean crushers, reliant on American soybeans and concerned that a closed market might benefit alternative oilseed crops. Therefore, the government put forward a policy for the protection of domestically produced soybeans, as a prerequisite for liberalization, which included: 1) increasing the import tariff from 10 percent to 13 percent; 2) providing a subsidy of 3 billion yen to domestic soybean producers and collection agents by the end of the 36th fiscal year; 3) enacting the "Temporary Measures Act for the Allocation of Soybean and Rapeseed Subsidies [*Daizu natane kōfukin zansei sochi hō*]" to determine the method of distributing the subsidy; 4) advancing research to rationalize the production of domestically grown soybeans.[277]

The measures to protect the national agricultural market were outlined in the Agricultural Basic Law (*Nōgyō kihon hō*) of June 12, 1961. The law established the guiding principles for Japanese agriculture, with the goal of preserving the vitality of rural areas and ensuring farm incomes at levels comparable to those in urban areas. To achieve these objectives, the law focused on improving the efficient use of natural resources, modernizing agricultural management, stabilizing agricultural input prices, developing human capital, and investing in rural areas.[278] In practice, however, the law did not provide the conditions for reducing dependence on soybean imports from abroad. Instead, it laid the foundation for the development of a dual-structure agricultural economy. Through selective expansion of agricultural production, on one side, it promoted the production of livestock, vegetables, and fruits while, on the other, it increased the imports of feed grains and other commodities to be produced in the land-extensive farming, with the only exclusion of rice.[279] In this situation, with the enactment of the Agricultural Basic Law, the government also implemented the liberalization of soybean imports. Starting from July 1, 1961, Japan removed quota controls, thereby eliminating any limits on the quantities that could be imported. However,

---

[277] "Nōrinshō wa kyōkō no hara—daizu yunyū jiyūka—yushutsunyū [The Ministry of Agriculture, Forestry, and Fisheries is forcing through the liberalization of soybean imports—exports and imports]," *Asahi Shinbun*, 2 June 1961.

[278] Aurelia George Mulgan, *Japan's Agricultural Policy Regime*, Routledge, London, 2006, pp. 29–33.

[279] Hirasawa Akihiko, "Formation of Japan's food security policy: Relations with food situation and evolution of agricultural policies," Papers and Reports, Norinchukin Research Institute Co., Ltd., 2017, pp. 9–10.

the government also imposed a tariff increase from 10 percent to 13 percent as a measure to protect the domestic market.[280]

Despite the intentions, the tariff increase was not sufficient to protect domestic production from the surge in imports. The effects of liberalization were almost immediate: the area dedicated to soybean production decreased dramatically in just a few years, halving between 1960 and 1965,[281] while American soybean imports doubled in six years, from approximately one million tons imported in 1960 to over two million imported in 1966.[282] The case of soybeans, due to their significance in the Japanese diet and industrial uses, is undoubtedly the most emblematic example of the effects of liberalization in the 1960s. However, similar trajectories can also be observed for other commodities which were also liberalized. For example, imports of sorghum from the United States increased from a value of $7 million in 1960 to $119 million in 1967. Raisin imports reached $5.4 million in 1967 from virtually nothing in the 1950s, and lemon imports rose from $1.1 million in 1964, when they were liberalized, to $7.1 million in 1967.[283]

As we have already pointed out, these policies, on the one hand, perpetuated the dual structure of Japanese agriculture—strong protection and subsidies for rice farming, liberalization and dependence on foreign sources for other products. On the other hand, they were one of the key factors behind Japan's significant economic growth in the 1960s and its transformation into a consumer society, holding down workers' wage level by providing them with cheap food imported from abroad.[284] In this way, the Engel's coefficient, which denotes the percentage of one's income spent on purchasing food, declined from 44 percent in 1960 to 34 percent in 1970, a level similar to that of European countries.[285]

3.4 The 1970s food crisis and the "soybean embargo"

In 1956, the Economic White Paper declared the end of the postwar era by declaring "Mohaya sengo de wa nai [It is no longer postwar]," marking the dawn

---

280 "Shichigatsu kara daizu jiyūka [The liberalization of soybeans from July]," *Yomiuri Shinbun*, 16 June 1961.

281 See: <https://www.maff.go.jp/j/seisan/ryutu/daizu/d_data/attach/pdf/index-45.pdf>

282 William Brandemuhl, *Soybean Utilization in Japan*, p. 49.

283 Department of Agriculture, "Hearings Before a Subcommittee of the Committee on Appropriation," House of Representative, Ninety-first Congress, First Session, Washington, 1969, p. 254.

284 Hirasawa, "Formation of Japan's food security policy," p. 11.

285 Tsurumi Shunsuke, *A Cultural History of Postwar Japan 1945–1980*, London & New York: Routledge, 2011, p. 93.

of the great economic growth era in Japan. This emblematic phrase remained etched in the nation's collective memory, symbolizing the promise of a prosperous future.[286] And the future was indeed bright. The decade of the 1960s was not only characterized by widespread prosperity ("the Golden Sixties") but also marked Japan's reintegration into the community of advanced nation as exemplified by significant international events like the Tokyo Olympics, Kawabata Yasunari's Nobel Prize, and the Osaka World Expo. Japan's economy had grown to the point where it even surpassed West Germany in size, making Japan the second largest economy in the world. From the point of view of food, Japan had entered an era of food abundance (*hōshoku jidai*), making the memories of food shortages from just two decades earlier a distant past. During this time, the Japanese diet evolved with new culinary habits that came from the other side of the ocean, but it also led to increased reliance on the United States for agrifood imports. Consequently, Japan's food self-sufficiency rate, calculated on a caloric base, started to decline from 75 percent in 1961 to 49 percent by 1970.[287] Nevertheless, with the exception of some voices within the M.A.F.F.,[288] this reliance was not seen as a problematic issue, due to the reciprocal advantages stemming from the economic ties with the United States, as discussed earlier, and the prevailing optimism driven by the increasing prosperity. However, in the early 1970s, the situation began to change, marking the onset of a series of crises that significantly impacted international trade and also strained the relationship between Japan and the United States. The 1973 oil crisis and the collapse of the Bretton Woods system caused economic

---

286 Shimizu Kazuhiko. "Mohaya 'sengo' de wa nai to iu shakaiteki kioku no kōsei katei" [The compositional process of the social memory that "it is no longer the 'postwar'"], in *Edogawa Daigaku kiyō* [Edogawa University bulletin], 25 (March) 2015, pp. 195–206.

287 Policy Division of the Ministry of Agriculture, Forestry, and Fisheries of Japan, "Food Supply and Demand Table–2022 [*Shokuryō jukyū hyō—Reiwa 4*]," pp. 15–17. Full text available at: <https://www.maff.go.jp/j/zyukyu/fbs/attach/pdf/index-20.pdf>

288 Among the most critical opinions toward dependence on the United States was Ōgura Takekazu, chairman of the Agriculture and Fisheries Research Council, an official arm of the Ministry of Agriculture and Forestry which directs the government's technical and economic research programs in agriculture. In 1966, he wrote: "We must consider whether it is advisable for the nation to remain dependent mainly on imports from the United States as at present. […] The total dependency on the United States, however, cannot be regarded as desirable, in view of the fact that the food importing countries of the world are becoming more and more dependent on the United States. It is necessary for Japan to have more diversified sources of food supply in order to secure a stabilized food supply." Ōgura believed that Japan should assist the agricultural sectors of Southeast Asian countries through investments and increase imports from that region. From: Joseph R. Barse, *Japan's Food Demand and 1985 Grain Imports Prospects*, U.S. Department of Agriculture, Washington D.C.: United States Government Printing Office, 1969, pp. 64–65.

disruption and lead to a period of uncertainty and instability. But, for the purposes of this work, we will focus on another crisis that occurred at the beginning of the 1970s, which was only partially related to the other two, that is the world food crisis. The 1970s food crisis marked the first major global food crisis since the end of World War II, resulting from a combination of factors, including climate change, poor harvests, inflation, and also diplomatic confrontations. The crisis's severity compelled the international community to convene the first World Food Conference at the F.A.O. headquarters in Rome in 1974, in order to address the issues related to food security, foster international cooperation, and alleviate the consequences of food shortages and escalating prices. Below, we will analyze the causes of the food crisis, with a particular focus on the international and diplomatic causes. Subsequently, in the following section, we will examine the impact of the crisis on food relations between the United States and Japan.

Between 1972 and 1974, there was a drastic surge in food prices on the global market, leading to widespread disorder in the international food system. In 1972, the output of food in the world declined for the first time in more than 20 years. In particular, world output of cereals fell by a significant amount of 33 million tons, creating a heavy deficit worldwide. As result, the wheat stocks of the main exporting countries fell from 49 million tons in 1971–72 to 29 million tons in 1972–73.[289] While the main cause behind the world food crisis was attributed mainly to extraordinary bad weather conditions in 1972, there was also a "most unusual coincidence of a number of events" which impacted the world food production and food prices.[290] One of these was the significant increase of world food consumption, driven on one hand by population growth and the economic boom in developing countries, and on the other hand by the "meatification" of food consumption in industrialized countries, leading to an increase in feed grain cultivation.[291] Moreover, the rise in oil prices in 1973 not only increased the cost fuel in agriculture but also led to shortage and high prices of feed stocks for fertilizer manufacture.[292] But the causes of the world crisis must be sought also in the agricultural policies of the United States and their determination to reduce the surplus system that had characterized global food trade until that point. The shift in U.S. policy direction commenced in the late 1960s when, due to growing discontent with the substantial expenses associated with the high

[289] United Nations Food Conference 1974, *Assessment of the World Food Situation. Present and Future*, E_CONF.65/3, p. 1.
[290] Ibidem.
[291] Jennifer Clapp, *Food*, Cambridge: Polity, 2012, pp. 45–6.
[292] U.N. Food Conference, E_CONF.65/3, p. 2.

costs of food storage for the global benefit of maintaining stable food prices, the country adopted a deliberate approach to reduce its vast government-held surplus grain reserves.[293] The change in perspective on the issue was largely driven by the ideological shift of U.S. President Nixon, who aimed to decrease government involvement in farm business and increasingly rely on free enterprise. This change was reinforced by the 1973 farm legislation, establishing that the U.S. government would no longer accumulate grain stocks.[294] The monetary policy pursued by Nixon, marked by the end of the Bretton Woods system and the establishment of a regime based on flexible exchange rates, also influenced agrifood trade. The depreciation of the U.S. dollar, the currency in which international prices of agricultural products were denominated, made American exports even more attractive on the international market.[295] In this context, the Soviet Union, also grappling with poor harvests due to drought, deemed it necessary to seek assistance from the United States and import agricultural products. Taking advantage of the ongoing process of détente, Richard Nixon and Henry Kissinger strategically utilized agricultural exports as a pivotal tool to advance both domestic farming and foreign policy objectives while forging a novel relationship between the United States and the Soviet Union. In July and August 1972 six American export companies entered into contracts with Exportkhleb, the Russian trading agency, to vend over 400 million bushels (over 1 million metric tons) of U.S. wheat valued at about 700 million dollars. Each contract was fixed price, calling for shipment of U.S. wheat from the 1972–73 crop.[296] To facilitate the sale of U.S. grains, the United States and the Soviet Union signed on July 8, 1972, "the largest long-term commercial trade purchase agreement ever made between two countries" and the U.S.S.R. agreed to buy a total of $750 million worth of American grain during the three-year period beginning August 1, 1972.[297] Secretary of State Henry Kissinger saw this agreement as "a major step forward in the conclusion of more comprehensive arrangements in other fields as well" and claimed that in the future the "general relationship with the Soviet Union would improve."[298] With the new agreement, the Soviet Union would have become the second largest

---

[293] Clapp, *Food*, pp. 48–49.
[294] Ibidem.
[295] Ibidem.
[296] U.S. Department of Agriculture, "Exporter's profits on sales of U.S. wheat to Russia," February 12,1974, Govern Accountability Office, B-176943, p. 6.
[297] U.S. Government, *Weekly Compilation of Presidential Documents*, vol. 8, n. 29, July 17, 1972, p. 1142.
[298] Ibidem, p. 1144.

purchaser of United States grain products after Japan.[299] However, the hope for a more comprehensive collaboration quickly faded away and the agreement turned to be a "colossal fiasco."[300] A month later, experts from the U.S. Department of Agriculture (U.S.D.A.) discovered the Soviet Union had purchased its full three-year $750 million grain allotment, using its credit to the fullest extent possible. Moreover, using its hard currency, the Soviet Union also purchased nearly $500 million of additional grain.[301] In a short amount of time, the U.S.S.R. had bought one-fourth of the 1972–73 wheat crop and large quantities of corn and soybeans, a quantity that surprised U.S.D.A. officials who had miscalculated Soviet needs.[302] The magnitude of this transaction precipitated an abrupt and unparalleled shortage, exacerbated by poor weather conditions worldwide and leading to a surge in prices. Despite the subsequent return of surpluses within a few years, attributed to the continuity of agricultural commodity programs that instigated them, the persisting tensions were exacerbated by factors such as agricultural and state debt, global competition, and shifts in the geopolitical power dynamics among nations.[303] The consequences were devastating for American economy as the surplus shortage sent grain prices soaring, pushing the inflation up. For this reason, the U.S. government was to pay large grain companies $333 million in export subsidies and over $46 million in shipping subsidies to help move the grain to the U.S.S.R.[304] This deal, from now on remembered with the name of "the Great Grain Robbery," marked the end of the postwar system of food trade based on the American surpluses.[305]

In an attempt to control the rise in food prices, President Nixon declared that "a new system for export controls" was necessary to restrain "the price of animal feedstuffs and other grains in the American market."[306] For this reason, the Nixon administration imposed several export embargoes on agricultural goods

---

299 Philip Shabecoff, "Moscow agrees to buy U.S. grain for $750 million," *The New York Times*, July 9, 1972.

300 Roger B. Porter, *The U.S.-U.S.S.R. Grain Agreement*, Cambridge, Cambridge University Press, 1984, p. 6.

301 Ibidem.

302 Mark J. Penn, "America Gets the Shaft," The Harvard Crimson, 16 November 1973. <https://www.thecrimson.com/article/1973/11/16/america-gets-the-shaft-pthe-1972/>.

303 Harriet Friedmann, "The political economy of food: a global crisis," in *New left review*, vol. 197, n. 29, p. 40.

304 Roger B. Porter, *The U.S.-U.S.S.R. Grain Agreement*, Cambridge, Cambridge University Press, 1984, p. 6.

305 Friedmann, "The political economy of food," p. 99.

306 Cited in: I. M. Destler, "United States Food Policy 1972–1976: Reconciling Domestic and International Objectives," in *International Organization*, vol. 32, n. 3, p. 628.

between 1973 and 1975. In particular, on June 27, 1973, Secretary of commerce Fredrick B. Dent, with the approval of the Secretary of Agriculture Earl L. Butz, declared "the most dramatic food policy decision of 1973": a complete, temporary embargo on American exports of soybeans, cottonseeds, and other agricultural products.[307] The government's primary goal in making this decision was initially to control domestic food prices. However, despite its short-lived nature, the decision had significant adverse effects on American foreign policy. Referred to as the "soybean embargo," this action "challenged the credibility of the U.S. commitment to free trade among nations and its reputation as a reliable supplier." Compounding the issue, the government's failure to effectively communicate its intentions to impose restrictions contributed to the development of skepticism and uncertainties among European and Japanese allies.[308]

3.5 The impact of the crisis on Japan

"The food crisis used to depend on the weather, but lately, it has become dependent on Nixon." With these words, the Yomiuri Shinbun attacked the Nixon administration in the aftermath of the U.S. decision to restrict agricultural and food exports.[309] The "soybean shock [*daizu shokku*]," as it was called by the Japanese press,[310] was just the latest in a series of decisions made by the Nixon administration that had the effect of straining the relationship between the United States and Japan, after the end of the Bretton Woods system in and the rapprochement with Beijing in 1971. However, the impression this time was that the American decision was primarily directed toward Japan, which was the foremost (and perhaps sole) true victim of the embargo, given that it relied on soybean imports for 91 percent of its needs and represented 15 percent of American soybean exports

---

307 Ibidem, p. 627.

308 U.S. General Accounting Office, "U.S. Actions Needed to Cope with Commodity Shortages," Report to the Congress, April 29, 1974, B-114824.

309 Citation taken from: Hoshina Teruhiko, "Shokuryō jikyū minaosu toki. Keizai bēsu hanarete [Reconsidering food self-sufficiency. Moving away from an economic base]," Yomiuri Shinbun, 29 June 1973.

310 See for example: "Daizu shokku, sakana ni tsuzuki shokutaku ni nayami, tōfu mo abura mo nattō mo—kome no daizu-rui yushutsu teishi [The soybean shock, following the fish, brings concerns to the dining table—tofu, oil, natto—suspension of soybean exports from the United States]," *Asahi shinbun*, 28 June 1973; "Kondo wa daizu shokku! Tōfu, miso ni eikyō [Now it's the soybean shock! Impact on tofu and miso]," *Yomiuri Shinbun*, 28 June 1973.

in 1971.[311] It was not a political shock but a "household shock" as it had "greater impact upon the ordinary Japanese citizen than on the Tanaka government."[312] Moreover, although the American intention to restrict soybean exports had been announced earlier, the ban was announced at a date "much earlier than expected" and its contents were "very strict."[313]

The Minister of Agriculture, Sakurauchi Yoshio, immediately called the American ambassador in Tokyo, Robert Ingersoll, to express his disappointment and to seek a remedy. Sakurauchi stated that the sudden announcement of the embargo was "shocking news" for Japan and a source of great confusion, not only because soybean foods were a "necessity of life" in Japan but also because "the threatened impact of the embargo on these vital commodities [would have] serious political implications."[314] Sakurauchi also said that the soybean issue was most worrisome because it revived fears about U.S.-Japan trade problems that might seriously damage relations between the two countries. For this reason, he requested Japan be permitted to receive all of the 660 thousand tons of soybeans on contract for shipment to Japan between July and September, in order to prevent the situation from worsening.[315] On his part, Ambassador Ingersoll immediately sought to reassure Sakurauchi by stating that the measures taken by the United States were aimed at protecting not only American consumers but also traditional foreign customers, such as Japan, who was their "best traditional overseas cash customer for agricultural products."[316] Simultaneously, Sakurauchi announced the decision to dispatch a survey mission to the United States to investigate Washington's motivations behind the export ban on soybeans and cottonseed. Led by Watanabe Mamoru, a councilor at the ministry secretariat, the team aimed to appeal to the U.S. government for measures ensuring a stable supply of agricultural products to meet Japan's needs. In parallel efforts to diversify sources of agricultural imports and reduce reliance on the United States, Foreign Minister Ōhira Masayoshi directed ministry officials to explore ways of promoting

311 United States International Trade Commission, *U.S. Embargoes on Agricultural Exports: Implications for the U.S. Agricultural Industry and U.S. Exports*, Report on Investigation No. 332-157 Under Section 332 of the Tariff Act of 1930, Washington D.C., December 1983, p. 5.

312 Hirazawa Kazushige, "Japan's Soybean Shock," *The Japan Times*, 6 July 1973.

313 "U.S. Slaps Immediate Embargo on Soybean, Cotton Seed Exports," *The Japan Times*, 29 June 1973.

314 U.S. Department of State, EO Systematic Review, "Visit with Japan Minister of Agriculture—Soybean Embargo," Document Number 1973TOKYO08249, 2005. URL: <https://aad.archives.gov/>

315 Ibidem.

316 Ibidem.

agriculture in Southeast Asian countries and acquiring products resulting from development assistance.[317]

The embargo lasted only for a brief period from June 27 to July 2, and it was quickly replaced by an export control system active until October 1 of the same year, which allowed exporters to ship 50 percent of their soybeans. From a purely agricultural and economic standpoint, the embargo did not induce any alterations in Japanese food imports. However, the deterioration in relation with the United States was predominantly psychological and political in nature. The soybean embargo undermined the trust that had been built up in the postwar period and had clearly demonstrated the fragility of a system built on heavy dependence on a single supplier. During the discussions at the U.S.-Japan Ministerial Committee on Trade and Economic Affairs held in Tokyo from July 16 to 17, 1973, Japan raised serious questions as to much it could rely on the United States as a future supplier. Criticisms mounted against the existing policy, placing the government under significant pressure to reduce reliance on the U.S. and diversify its food supplies. Beyond agricultural considerations, some observers contended that the perceived unreliability in the agricultural domain reflected broader concerns about the United States' consistency in political and security matters. There was a growing perception of a nationalistic America willing to compromise international rules and friendships for its strategic interests, raising fundamental questions about the stability of diplomatic relations.[318] Japanese Prime Minister Tanaka Kakuei's visit to Washington from July 31 to August 1 was an attempt to ease tensions between the two nations. Nixon referred to Japan as an "equal partner," supported Japan's attempt to become a permanent member of the United Nations Security Council, expressed satisfaction with Tanaka's achievements in reducing Japan's trade surplus with the United States and pledged his "best efforts" to assure Japan of sufficient food supplies.[319] Regarding the soybean embargo, he acknowledged that it "might have been good for domestic politics" but it was "disastrous" in terms of foreign policy as it seriously damaged relationship with Japan, especially considering prior efforts to persuade Japan to increase its purchase

---

317 "Japan's Group to Study U.S. Soybean Situation," *The Japan Times, 29 June 1973.*

318 "Memorandum From Robert Hormats of the National Security Council staff to the President's Assistant for National Security Affairs (Kissinger)," in *Foreign Relations of The United States, 1969–1976, Volume E–12, Documents on East and Southeast Asia, 1973–1976*, Document 177, Washington, July 19, 1973. URL: <https://history.state.gov/historicaldocuments/frus1969-76ve12/d177>

319 Richard Halloran, "Visitor Tanaka," *The New York Times*, 5 August 1973.

of agricultural products.[320] Nevertheless, although all restrictive measures were soon withdrawn, the issue of the soybean embargo persisted, characterizing the diplomatic debate between Japan and United States for many months. During his visit in Tokyo for the American Food Festival from April 13 to 18, 1974, Secretary of Agriculture Earl Butz openly acknowledged that the soybean embargo "was a mistake" but it taught the United States that they "should never impose an embargo again."[321] Prime Minister Tanaka stressed Japan's view that food was "a more critical resource than fuel" and proposed the establishment of more frequent ad hoc consultations between experts from both nations to better address food-related concerns.[322]

The commitment to ensuring a consistent soybean supply was once again emphasized also during Tanaka's visit to Washington in the subsequent September. Because of another unsatisfactory harvest, there were concerns about Japanese supplies. President Ford, who took office after Nixon's Watergate scandal, reassured that the United States would not impose any export restrictions.[323] During his historic visit to Japan in November, the first time for a sitting President, Ford reiterated the U.S. efforts to "continue as a big supplier of an adequate food supply for Japan" and assured that his administration was not "going to take arbitrary action to cut off agricultural exports."[324]

The prevailing perception was that the United States had become fully aware of its diplomatic misstep and was actively seeking to mitigate its impact by reassuring its ally through formal expressions of apology and support. Yet, despite official reassurances, a certain tension continued to persist behind the scenes. Japan sought assurance regarding stockpiling due to concerns about the reliability of the United States as a supplier. However, this assurance was not

320 "Memorandum of Conversation," in *Foreign Relations of The United States, 1969–1976, Volume E–12, Documents on East and Southeast Asia, 1973–1976*, Document 179, Washington, July 31, 1973. URL: <https://history.state.gov/historicaldocuments/frus1969-76ve12/d179

321 U.S. Department of State, EO Systematic Review, "Summary of Secretary Butz Visit-April 13–18, 1974 Begin Summary: During Five-Day Visit to Japan Secretary Butz," Document Number 1974TOKYO05373. URL: <https://aad.archive.gov/>

322 Ibidem.

323 "Memorandum of Conversation," in *Foreign Relations of The United States, 1969–1976, Volume E–12, Documents on East and Southeast Asia, 1973–1976*, Document 179, Washington, September 21, 1974. URL: <https://history.state.gov/historicaldocuments/frus1969-76ve12/d195>

324 "Memorandum of Conversation," in *Foreign Relations of The United States, 1969–1976, Volume E–12, Documents on East and Southeast Asia, 1973–1976*, Document 198, Tokyo, November 19, 1974. URL: <https://history.state.gov/historicaldocuments/frus1969-76ve12/d198>

attainable under the existing preferential treatment that Japan already enjoyed.[325] In this context, an agreement leading to medium-term supply and purchase commitment was deemed the best solution to restore trust between the two countries. The agreement was reached on August 12, 1975, during the visit of the Japanese Minister of Agriculture, Abe Shintarō, to Washington D.C. with the U.S. Secretary of Agriculture, Earl Butz. The outcome of this meeting was the Abe-Butz Gentlemen's Agreement, described "not as a detailed treaty" but as a "mutual confirmation of the intentions of the two governments."[326] The agreement stipulated that Japan would annually purchase 14 million tons or more of American cereals and soybeans, including 3 million metric tons of wheat, 3 of soybeans, and 8 of feed grains, for the next three years.[327] Recognizing that "the United States [was] by far the most important market for Japan in terms of the export of its industrial products, and likewise, Japan [was] the most crucial market for the United States in terms of the export of its agricultural products,"[328] Butz assured that "supplies to Japan would not be cut off by export controls, such as the temporary restrictions on soybean exports, which strained Japanese-American trade relations in 1973."[329] In this way, the Butz-Abe Gentlemen's Agreement served as the tangible attempt to restore Japan's confidence in United States' reliability as a supplier of agricultural products.

3.6 Resource diplomacy and comprehensive security

The occurrence of the food crisis, the "soybean embargo," and the related concerns over using food as a diplomatic weapon instigated strong political pressures in Japan to reorient its long-term food policies toward increased domestic production and the exploration of alternative import sources. As early as the summer of 1972, at the onset of the food crisis and even before the soybean embargo, M.A.F.F. had requested the Ministry of Finance to propose the establishment of a new agency. This agency would be tasked with the development of the agricultural sector

---

325 "Minutes of Senior Review Group Meeting," in *Foreign Relations of The United States, 1969–1976, Volume E–12, Documents on East and Southeast Asia, 1973–1976*, Document 197, Washington, November 11, 1974. URL: <https://history.state.gov/historicaldocuments/frus1969-76ve12/d197>

326 "U.S.-Japan Grain Soybean Agreement Announced in D.C.", *The Japan Times*, 14 August 1975.

327 "Nichibei nōsō. Kokumotsu yunyū mokuhyō de gōi [Japan-U.S. Agricultural Ministerial Meeting. Agreement on Grain Import Targets]", *Yomiuri Shinbun*, 13 August 1975.

328 "Nichibei nōsō no kyōdō happyō [Joint communiqué of Japan-U.S. Ministries of Agriculture]", *Yomiuri Shinbun*, 13 August 1975.

329 "Japan Arranges to Buy U.S. Grain," *The New York Times*, 13 August 1975.

in developing countries, based on the principle of "kaihatsu-yunyū (develop-and-imports)."[330] The concept emerged in the 1950s within the framework of Japan's definition of "economic cooperation," involving capital, technology, and trade. Initially applied to promote exports of Japanese plant and equipment, it later expanded to include primary products, including agrifood products. By the late 1960s, "kaihatsu-yunyū" became a key element in Japan's concern for the stability of raw material supplies. In the context of agricultural commodities, the aid provided was central to Japanese efforts to exert more control over its sources of import.[331] Through the establishment of the new agency, M.A.F.F. officials hoped to increase investments abroad in the agricultural sector of developing countries—which accounted only for 4.1 percent of Japanese projects abroad—in order to enhance the diversification of import sources.[332] However, initially, the objectives of M.A.F.F. were hindered by internal opposition from other ministries, particularly from M.I.T.I., which sought a similar allocation of resources for the Small and Medium Industry Overseas Investment Agency. Apparently, the initial inclination of the Ministry of Finance was in favor of M.I.T.I.'s project, as past initiatives in the agricultural development had not had any real impact on Japan's food imports. However, the soybean embargo had a decisive effect on the debate, shifting the advantage in favor of the M.A.F.F.'s proposal. The once prevalent apathy of 1972 was replaced by intense political involvement and lobbying by proponents of M.A.F.F. ideas, garnering support from politicians. Subsequently, the decision to formally establish the new Japan International Cooperation Agency (J.I.C.A.) was reached during the Budget Cabinet meeting on the morning of Saturday, December 29, 1973. The J.I.C.A. bill was ratified by the Cabinet on February 15, 1974, presented to the Diet on February 18, and successfully passed through the House of Representatives on May 14, following deliberations in the Foreign Affairs Committee. The House of Councilors approved it on May 27, and the bill was officially promulgated as Law No. 62 on May 31, 1974.[333]

Concurrently with the establishment of J.I.C.A., the 1970s raw material crises (food and energetic) compelled the then-Japanese Prime Minister, Tanaka Kakuei, to pursue a more proactive foreign policy, going beyond the framework of bilateralism with the United States. From the point of view of agricultural and food

330 Alan Rix, *Japan's Economic Aid. Policymaking and Politics*, London: Routledge, 2011, p. 30.
331 Alan Rix, "The Mitsugoro Project: Japanese Aid Policy and Indonesia," in *Pacific Affairs*, 1979, vol. 52, n. 1, pp. 51–52.
332 Rix, *Japan's Economic Aid*, p. 30.
333 Ibidem, pp. 30–50.

imports, Japan launched an independent and vigorous campaign of "resource diplomacy (*shigen gaikō*)," as a way to re-establish its access to these vital supplies.[334] The Japanese government thus decided to launch a food diversification strategy, encouraging joint public-private ventures to consolidate alternative food sources. By means of this strategy, the Japanese government intended to utilize Official Development Assistance (O.D.A.) funds through J.I.C.A. to stimulate agricultural production in developing nations. Concurrently, through the provision of tax incentives, efforts were made to facilitate Japanese agribusiness corporations' investments in these countries with the goal of producing food for importation to Japan. M.A.F.F. initially proposed investments in Southeast Asia and Africa, specifically targeting Indonesia and Madagascar due to their extensive untapped land for potential food production. However, the plans faced setbacks in 1974 when Prime Minister Tanaka's visit to Jakarta sparked violent anti-Japanese protests, resulting in a reevaluation and reduction of Japan's agricultural development goals. Simultaneously, it became apparent that African nations lacked essential infrastructure for launching comprehensive agricultural development initiatives on a significant scale.[335]

One of the solutions to Japan's excessive dependence on the United States was found in Brazil. In 1973, the Ministry of Agriculture organized an expedition led by Matsumoto Sakuei, the head of M.A.F.F.'s Economic Cooperation Division, accompanied by four other specialists. Their objective was to investigate the potential of utilizing the uncultivated region of Cerrado, a vast tropical savanna ecoregion in the middle of Brazil, for the large-scale production of essential agricultural products to export to Japan, in particular soybeans.[336] The results of the expedition were favorable, and in September 1974, the Prime Minister of Japan, Tanaka Kakuei, and the President of Brazil, Ernesto Geisel, announced in a joint communiqué their commitment to technical and financial collaboration for the agricultural development of Brazil.[337] However, several years of technical studies and scientific research were required before a comprehensive Cerrado development project could be implemented.

---

334 Oliviero Frattolillo, *Il Giappone tra Est e Ovest. La ricerca di un ruolo internazionale nell'era bipolare*, Milano: FrancoAngeli, 2014, pp. 150–156.

335 Okada Kana, "Role of Japan in Offshore Agricultural Investment. Case of ProSAVANA Project in Mozambico," Research Paper, International Institute of Social Studies, The Hague, 2014, p. 17.

336 Hosono Akio, Carlos Magno Campos da Rocha and Hongo Yutaka, *Development for Sustainable Agriculture. The Brazilian Cerrado*, New York: Palgrave MacMillan, 2016, p. 222.

337 The text of the joint communiqué is available at: M.O.F.A., *Bluebook 1975*, url: <https://www.mofa.go.jp/mofaj/gaiko/bluebook/1975_2/s50-contents-4.htm>

In February 1975, J.I.C.A. dispatched a technical mission to Brazil, led by Ashikaga Tomomi, Director of the Japanese Department of Agricultural and Forestry Planning and Research, to explore potential Japanese-Brazilian collaboration in agriculture. The same year saw the establishment of the Commission for the Development of Japanese-Brazilian Agricultural Cooperation by Keidanren, involving representatives from the industrial sector, M.A.F.F., M.O.F.A., and J.I.C.A. Furthermore, in February 1976, a group led by Hisamune Takashi, Deputy Director of J.I.C.A., engaged in a series of meetings with representatives from the Brazilian government and the private sector to discuss the initial version of the cooperation program.[338] Concurrently, Brazil introduced the "Program for the Development of the Cerrado (Programa de Desenvolvimento dos Cerrados)" in 1975, aiming to enhance economic and social infrastructure necessary for exploiting the area. To address agricultural studies and research, the Cerrado Agricultural Research Center (Centro de Pesquisa Agropecuária dos Cerrados or C.P.A.C.) was established in the same year near Brasilia.[339] J.I.C.A. launched its technical cooperation program with C.P.A.C. in 1977, providing financial support, human capital, and equipment.[340] In 1979, the two governments officially launched the Cerrado Development Program, the culmination of the extensive period of study and research. Spanning 22 years (1979–2001), PRO.DE.CER. (Projeto de Desenvolvimento do Cerrado), how it was officially named, transformed over three hundred thousand hectares of the Cerrado into agricultural land primarily for soybean cultivation. Japan contributed over 28 billion yen to PRO.DE.CER. as development aid.[341] The program progressed through three main phases, during which Brazil's soybean production increased to levels comparable to the United States, allowing Japan to reduce its dependence on American imports.

The food crisis in the 1970s not only pushed the Japanese government to adopt a more independent and proactive foreign policy but also had a profound impact on the redefinition of the very concept of national security in Japan. While the Butz-Abe agreement aimed to reassure the Japanese concerning material food supplies, as highlighted earlier, the Japanese government's apprehensions extended beyond mere economic considerations. There were also concerns about the potential use of food as a political and diplomatic weapon. In the 1970s, the

[338] Akio Hosono, et al., *Development for Sustainable Agriculture. The Brazilian Cerrado*, pp. 251–252.

[339] Ibidem, p. 55.

[340] Ibidem, 157–96.

[341] See: <http://www.maff.go.jp/j/kokusai/kokkyo/toushi/pdf/1304mgj4.pdf>

United States became more conscious of the significant power derived from being the world's primary food producer, and the idea of utilizing food resources as a diplomatic weapon began to take root, akin to the role played by oil or other commodities. In August 1974, a month before the World Food Conference held in Rome, the C.I.A. issued a report where it was concluded that "the U.S.'s near monopoly position as food exporter would have an enormous, though not easily definable, impact on international relations [and] (i)t could give the U.S. a measure of power it had never had before."[342] In November of the same year, the Secretary of Agriculture, Earl Butz more emphatically declared: "Food is a weapon. It is now one of the principal tools in our negotiating kit."[343] The idea of using food not just as a means to support allies but also as a punitive tool against adversaries influenced foreign policy discussions from the 1970s until the early 1980s. This emphasis on the "food weapon" persisted until the Carter administration's embargo against the Soviet Union in 1980, issued in response to the invasion of Afghanistan, demonstrated its ineffectiveness.[344]

The idea that economic resources such as oil, food, and other raw materials could serve as diplomatic weapons to disrupt the security of nations relying on imports prompted Japan to adopt a new perspective on security. This evolved concept encompassed not only "traditional" military threats but also acknowledged "non-traditional" threats, specifically those of an economic nature. In 1972, the Japan Economic Research Institute, in collaboration with the Economic Affairs Bureau of the M.O.F.A., published a report on Japan's dependence on foreign sources of raw materials, emphasizing the necessity for the Japanese government to implement a "resource diplomacy" to ensure a stable acquisition of these materials and uphold national security. The report stated that "in an era marked by intense competition for resource acquisition, resource scarcity is a critical issue directly linked to National Security (in English and capitalized in the original text), often carrying significant military implications."[345] The term "economic security (*keizai anzen hoshō*)" started to be widely used to define the ultimate

[342] It is interesting to note that the conclusions of the report were based on forecasts of cooler global temperatures in the future. See: Central Intelligence Agency, *Potential Implications of Trends in World Population, Food Production, and Climate*, August 1974, p. 39.

[343] Henry Weinstein, "Big U.S. Grain 'Weapon'," *The New York Times*, 17 March 1975.

[344] Robert L. Paarlberg, "Lessons of the grain embargo," in *Foreign Affairs*, vol. 59, n. 1, 1980, pp. 144–62.

[345] Japan Economic Research Institute, Economic Affairs Bureau of the Ministry of Foreign Affairs, *70 nendai ni okeru shigen gaikō* [Resource diplomacy in the Seventies], Ministry of Finance Press, Tokyo, 1972, p. 3.

goal of Tanaka's resource diplomacy.[346] But it was in 1977 that a new concept was created, bridging traditional security and economic security: comprehensive security (*sōgō anzen hoshō*). The Nomura Research Institute (N.R.I., or *Nomura sōgō kenkyūsho* in Japanese), established in 1965 with a focus on policy advising and corporate strategy development, officially introduced the term "comprehensive security" in its 1977 report titled "Kokusai kankyō no henka to Nihon no taiō (Changes in the International Environment and Japan's Response)."[347] In the report, the comprehensive security was considered as a combination of the concepts of military security and economic security. Recommendations included augmenting the budget for essential raw material imports, including agricultural products, establishing material reserves, and investing in new technologies to bolster national production.[348] The formulation of a comprehensive security strategy became a focal point during the Ōhira administration (1978–1979). In 1979, Ōhira established the Group on the Research of Comprehensive Security (*Sōgō anzen hoshō kenkyū gurūpu*), which produced in 1980 the Report on the national comprehensive security (*Sōgō anzen hoshō kenkyū gurūpu hōkoku*) where they identified six main goals for Japan, among which food security.[349] For what concerns food security, according to the report, the main threats to Japan's food security were to be found in poor harvests in producer countries, deteriorating diplomatic relations, interruptions in sea routes, and demographic imbalances in comparison to food production. The report outlined Japan's strategy, emphasizing the enhancement of national self-sufficiency, further liberalization of agricultural products, and investments in developing countries abroad.

3.7 Conclusion

In the early 1960s, following the renewal of the Security Treaty, Japan entered a new phase often referred to as the "economic season (*keizai jidai*)." During this

346 Li Enmin, "Nihon no keizai anzen hoshō wo motomete. 1970 nendai ni okeru keidanren no shigen gaikō" [In Pursuit of Japan's Economic Security: Keidanren's Resource Diplomacy toward China during the 1970s], in *Kenkyū Ronshū*, vol. 8, 1999, pp. 67–116.

347 Nomura Research Institute, *Kokusai kankyō no henka to Nihon no taiō* [Changes in the International Environment and Japan's Response]. Tokyo: Sōgō kenkyū kaihatsu kikō, 1977.

348 Ibidem, p. 123–149.

349 The goals were: strengthening of the cooperation with the United States and the other liberal states; strengthening of the self-defense force of Japan; a change in the attitude toward U.R.S.S.; improvement of energy security; improvement of food security; improvement of the management of natural disasters. The full text of the report is available at: <https://worldjpn.net/documents/texts/JPSC/19800702.O1J.html>

period, the nation directed its efforts toward achieving high economic growth and improving the well-being of the Japanese, as exemplified by Prime Minister Ikeda's "double income plan." However, this ambitious pursuit of economic prosperity attracted strong criticism from the international community, who increasingly pressured Japan to adhere to international trade rules and align its economic system with established global standards. In this context, Ikeda's economic diplomacy was strategically designed to achieve the withdrawal of G.A.T.T. Article 35 reservation, get O.E.C.D. membership, and entry into the economic order and club of Western advanced nations while simultaneously reorganizing the domestic industrial and economic structure. From the perspective of agrifood imports, adherence to international rules through a liberalization program led to a significant increase in imports from abroad. On the one hand, this had a detrimental effect on domestic production; however, on the other hand, it ensured access to cheap food, thereby supporting economic growth during those years.

The food crisis of the 1970s and the subsequent "soybean embargo" imposed by the United States instilled a profound sense of mistrust toward the American ally. Simultaneously, it prompted Japan to reevaluate the concept of national security and the objectives of its foreign policy. The formalization of the concept of comprehensive security, within which food security constituted a pivotal element, combined with resource diplomacy, enabled Japan to dismantle the structure based on its relationship with the United States, that had guaranteed the stability of food supplies since the early postwar years. Through investments and aids in developing countries, Japan sought to diversify its food supplies, thereby gaining direct control over its food imports from abroad. This strategic shift marked a pivotal moment in Japan's postwar trajectory, reflecting a deliberate move toward reduced dependence on the United States and fueling a debate on national self-sufficiency that has become central in all discussions on food security and continues to endure to this day.

Conclusions

In 2022, the United States accounted for approximately 18 percent of Japan's total food imports, rising to 23 percent when focusing exclusively on agricultural products.[350] While these figures confirm the United States as Japan's largest single food supplier, they are far removed from the staggering 90 percent dependence observed in the immediate postwar period. Japan's efforts in the 1970s to diversify its sources of food supply were notably successful in reducing reliance on American imports; however, its overall dependence on external suppliers remains acute. According to M.A.F.F., in 2023, Japan recorded a self-sufficiency rate in terms of calorie intake of 38 percent, one of the lowest in the world. Despite Japan being a country where food waste is a more pressing issue today, and food shortages seem like a distant memory from history books, the strong dependence on foreign sources keeps the "food problem"—expression usually replaced by "food security" [*shokuryō anzen hoshō*] since the 1970s—remains a pressing concern for Japan as it could potentially leave the country without food in the event of wars or trade disputes. For this reason, the "food problem" is still at the forefront of Tokyo's political and diplomatic agenda.

From a domestic policy perspective, Japan's heavy reliance on imports has sparked ongoing debates about promoting self-sufficiency by revitalizing national agricultural production and returning to traditional dietary practices. However, such initiatives often serve more as a rhetorical device to garner public support than as substantive policy objectives. In practice, even after the disruptions of the 1970s, Japan has continued to prioritize the import of cost-effective agricultural

[350] M.A.F.F., *Nōrin suisanbutsu yushutsunyū gaikyō* [Overview of Agricultural, Forestry, and Fisheries Import and Export], p. 9. Url: www.maff.go.jp/j/tokei/kouhyou/kokusai/attach/pdf/index-58.pdf

products, ensuring the stability of these imports while protecting access to international markets for its industrial exports. This pragmatic approach was exemplified in the prolonged negotiations for the liberalization of orange and beef imports. Despite strong pressure from the United States dating back to the early 1970s, Japan successfully resisted until 1988 when the agreement on the liberalization of oranges and beef was ultimately reached. However, Japan's acceptance of liberalization only materialized when the potential for retaliatory measures, particularly within the automotive industry, became apparent. Throughout the negotiations, Japan's representatives skillfully utilized their bargaining power to secure transition measures, resulting in an agreement that gradually liberalized the Japanese beef and orange markets on terms advantageous to Japan. This agreement involved raising quota ceilings annually for three years, transitioning to tariffs in 1991, and subsequent reductions in tariff rates. Similarly, for several years, Japan managed to withstand pressure from the United States to liberalize its rice market until 1993 when it decided to yield to American pressure and eliminate the ban on rice imports within the Uruguay Round negotiations. Once again, however, Japan prioritized maintaining an open international market for its industrial goods and succeeded in achieving only minimal liberalization of rice imports.[351]

In this book, we have attempted to demonstrate how this approach originated precisely in the postwar years, during the development of the relationship with the United States of America. We have seen how during the Occupation, the American decision to provide food aid to Japan, while undoubtedly beneficial for the disposal of surplus agricultural goods, was primarily motivated by genuine intentions to assist a nation brought to its knees by famine, as well as the strategic aim to halt the spread of revolutionary Communist ideas. These objectives were widely shared by the Japanese political élite, who saw American aid as the only possibility to begin alleviating the problem of hunger plaguing the population, to maintain social and political stability and to rebuild the country's economy. The food aid received during the post-Occupation years via programs like the M.S.A. or P.L.480 also witnessed proactive efforts from the Japanese government to secure them, driven by the objective of rejuvenating the industrial sector and conserving international currency reserves, as we have seen

[351] For details on the two negotiations see: Christina L. Davis, *Food Fights Over Free Trade. How International Institutions Promote Agricultural Trade Liberalization*, Princeton, Princeton University Press, 2003, pp. 113–175; Christopher C. Meyerson, *Domestic Politics and International Relations in U.S.-Japan Trade Policymaking. The G.A.T.T. Uruguay Round Agriculture Negotiations*, London, Palgrave McMillan, 2003, pp. 98–133.

in Chapter 2. Even when international pressure was at its peak in the late 1950s and early 1960s, demanding Japan to liberalize its agricultural market, the government's response to acquiesce to the demands of the international community was once again driven by the pursuit of both economic and international objectives. Economically, Japan obtained the removal of Article 35 of the G.A.T.T. by European countries, enabling it to import food at lower costs, thereby supporting the significant economic growth of those years. Internationally, it succeeded in fully reinstating itself as a member of the international community through its accession to the O.E.C.D. And finally, during the first major economic crisis of the postwar period, when the United States turned its back on its ally with the infamous "soybean embargo," Japan's response was once again to prioritize its own interests and implement a strategy aimed at reducing its dependence on Washington and gaining greater control over its food imports.

With this work, we tried to contrast the portrayal of Japan as having been passively subjected to the United States' "wheat strategy," a perspective often emphasized in the literature. In contrast to these prevailing assertions within the existing literature, this study endeavors to illustrate that Japan's dependence on foreign food sources, although often portrayed as a concern by M.A.F.F. or regarded as a threat to national food security, is fundamentally rooted in a strategic diplomatic approach adopted by Japan since the postwar years. Through this approach, born out of compromises among various forces, Japan has managed to advance its own economic and international interests, albeit at the expense of certain concessions. Importantly, Japan has effectively reframed this dependence from a perceived vulnerability into a mechanism through which it safeguards the stability and resilience of its food supply.

Bibliography

Primary Sources

Archives

Records of the U.S. Department of State Relating to the Internal Affairs of Japan, 1950–1954 (microform). Department of State decimal files 794, 894, and 994.

Digital Archives

Asahi Shinbun Digital Archive, URL: <https://xsearch.asahi.com/>

Japan Center for Asian Historical Records (JACAR), URL: <https://www.jacar.archives.go.jp/>

Japan' s national diet library digital collections, URL: <https://dl.ndl.go.jp/>

national archives of Japan digital archive, URL: <https://www.digital.archives.go.jp/>

Office of the historian, *The Foreign Relations of the United States*, Washington D.C., Department of State, URL: <https://history.state.gov/>

The Japan Times Digital Archives, Url: <http://jtimes.jp/archives>

The new york times archive, url: <https://timesmachine.nytimes.com/browser>

The U.S. National Archives and Records Administration, URL: <https://aad.archives.gov/>

World Trade Organization/Gatt documents online, Url: <https://docs.wto.org/>

Yomiuri database service, url: <https://database.yomiuri.co.jp/>

Reports

Barse, Joseph R. 1969. *Japan's Food Demand and 1985 Grain Imports Prospects*. U.S. Department of Agriculture. Washington D.C.: U.S. Government Printing Office.

Blakeslee, George Hubbard. 1953. *The Far Eastern Commission: A Study in International Cooperation: 1945 to 1952*. Washington D.C.: Department of State.

Central Intelligence Agency. 1948. *Strategic Importance of Japan*. Document number: 0000009376. Publication Date: April 1, 1948.

———. 1949. *The Strategic Importance of the Far East to the U.S. and U.S.S.R.* Document number: CIA-RDP78-01617A003400130008-2. Publication date: May 4, 1949.

———. 1974. *Potential Implications of Trends in World Population, Food Production, and Climate*. Document number: C00108382. Publication date: August 1974.

Committee on Foreign Affairs. 1980. *U.S. policy in the Far East, Part I, Vol XVIII*, Washington D.C.: U.S. Government Printing Office.

Galdi, Theodore W. 1977. *Use of U.S. Food Resources for Diplomatic Purposes: An Examination of the Issues*, Washington, D.C.: United States Government Printing Office.

Gray, Lewis C., Baker, Oliver E. 1930. *Land Utilization and the Farm Problem*. United States, Department of Agriculture.

Japan Economic Research Institute, Economic Affairs Bureau of the Ministry of Foreign Affairs. 1972. *70 nendai ni okeru shigen gaikō* (Resource diplomacy in the Seventies), Tokyo: Ministry of Finance Press.

Ministry of Health and Welfare (Kōseishō). 1996. *Lara kinenshi* [Lara memorial book]. Tōkyō: Kōseishō.

Nomura Research Institute. 1977. *Kokusai kankyō no henka to Nihon no taiō* [Changes in the International Environment and Japan's Response]. Tokyo: Sōgō kenkyū kaihatsu kikō.

Psychological Strategy Board. 1951. «Inventory of Cold War operations detailed». GALE|CK2349324991. United States: White House. October 11, 1951.

Sallagar, Frederick M. 1974. «Lessons from an Aerial Mining Campaign (Operation "Starvation")». Rand Report 1322-PR. Rand Report 1322-PR.

Supreme Commander for the Allied Powers. 1945a. «Summation No. 1 of Non-Military Activities in Japan and Korea». Tokyo: General Headquarters Supreme Commander for the Allied Powers.

———. 1945b. «Summation No. 3 of Non-Military Activities in Japan and Korea». Tokyo: General Headquarters Supreme Commander for the Allied Powers.

———. 1946. «Summation No. 5 of Non-Military Activities in Japan and Korea». Tokyo: General Headquarters Supreme Commander for the Allied Powers.

Supreme Commander for the Allied Powers—Natural Resources Section. 1951. «Fisheries Programs in Japan, 1945–51». Report no. 152. Tokyo: General

Headquarters, Supreme Commander for the Allied Powers. Natural Resources Section.

The Far Eastern Commission. 1947. «Activities of the Far Eastern Commission: Report by the Secretary General, February 26, 1946–July 10, 1947». Washington D.C.: Department of State, Division of Publications, Office of Public Affairs.

The United States Strategic Bombing Survey (USBS). 1947. *The Japanese Wartime Standard of Living and Utilization of Manpower.* Manpower, Food and Civilian Supplies Division.

United Nations Food Conference. 1974. *Assessment of the World Food Situation. Present and Future.* E_CONF.65/3

United States General Accounting Office. 1974. *U.S. Actions Needed to Cope with Commodity Shortages.* Report to the Congress, April 29, B-114824.

United States International Trade Commission. 1983. *U.S. Embargoes on Agricultural Exports: Implications for the U.S. Agricultural Industry and U.S. Exports.* Report on Investigation No. 332-157 Under Section 332 of the Tariff Act of 1930, Washington D.C., December 1983.

United States Pacific Fleet and Pacific Ocean Areas. 1944. «Psychological Warfare, Part 2. Propaganda Material». N-14996-B.

United States Senate. 1962. *Trade Expansion Act of 1962: Hearings Before the Committee on Finance. United States Senate, Eighty Seventh Congress, Second Session on H.R. 11970.* Washington D.C.: U.S. Government Printing Office, July 1962.

U.S. Army Air Force, A-3 20th Air Force. 1945. «Starvation. Phase Analysis of Strategic Mining Blockade of the Japanese Empire».

Vellianitis-Fidas, Amalia and Eileen Marsar Manfredi. December 1977. *P.L.480 Concessional sales—History, Procedures, Negotiating and Implementing Agreements.* Foreign Agricultural Economic Report No. 142. Foreign Demand and Competition Division, Economic Research Service, U.S. Department of Agriculture.

Diaries and memoirs

Dulles, John F. 1951. *A Peace Treaty in the Making*, Japanese Peace Conference, San Francisco, September 4–8, 1951, Washington D.C.: U.S. Government Printing Office.

Lansing, Robert. 1918. *Robert Lansing Papers: Private Memoranda, 1922.* Originals; 1918 to 1919. Manuscript/Mixed Material. <https://www.loc.gov/item/mss29454002/>.

MacArthur, Douglas. 1964. *Reminiscences.* New York: McGraw-Hill.

Yoshida, Shigeru. 1961. *The Yoshida Memoirs. The Story of Japan in Crisis*. Translated by Ken'ichi Yoshida. London: William Heinemann Ltd.

———. 1967. *Nihon wo ketteishita hyaku nen* [Japan's decisive century]. Tokyo: Nihon Keizai Shinbunsha.

Other

Arisawa, Hiromi and Inaba, Hidezō. 1966. *Shiryō. Sengo nijūnen-shi. Vol. 2, Keizai* [Documents. History of the twenty years after the war. Volume 2. Economy]. Tokyo: Nippon Hyōronsha,

Committee on Appropriation—United States Senate. 1949. «Hearings—First Session on H.R. 4830 A Bill making Appropriations for Foreign Aid for the Fiscal Year Ending June 30, 1950, and for other purposes». Washington D.C.: U.S. Government Print Office.

Department of Agriculture. 1969. «Hearings Before a Subcommittee of the Committee on Appropriation». House of Representative, Ninety-first Congress, First Session, Washington.

United States Government. 1972. *Weekly Compilation of Presidential Documents*, vol. 8, n. 29, July 17, 1972.

Secondary Sources (books, miscellaneous, scientific articles)

Aduard, E. J. Lewe van. 2013. *Japan: From Surrender to Peace*. Dordrecht: Springer Science & Business Media.

Assmann, Stephanie. 2015. «The remaking of a national cuisine: The food education campaign in Japan». In *The globalization of Asian cuisines: Transnational networks and culinary contact zones*, ed. James Farrer, 165–185. New York: Palgrave Macmillan U.S.

Auslin, Michael R. 2006. *Negotiating Imperialism. The Unequal Treaties and the Culture of Japanese Diplomacy*, Cambridge, Massachusetts: Harvard University Press.

Ball, W. Macmahon. 1948. «Reflections on Japan». *Pacific Affairs* 21 (1): 3–19.

Brandemuhl, William. 1965. *Soybean Utilization in Japan, Volume I*. Kyoto: University of Kyoto.

Bunker, Stephen G., and Paul S. Ciccantell. 2007. *East Asia and the global economy: Japan's ascent, with implications for China's future*. John Hopkins studies in globalization. Baltimore: Johns Hopkins University Press.

Clapp, Jennifer. 2012. *Food*. Cambridge: Polity.

COHEN, THEODORE. 1987. *Remaking Japan: The American Occupation as New Deal.* A cura di Herbert Passin. New York: Free Press.

COLLINGHAM, LIZZIE. 2011. *The Taste of War: World War Two and the Battle for Food.* London: Penguin U.K.

CWIERTKA, KATARZYNA J, 2006. *Modern Japanese Cuisine: Food, Power and National Identity.* London: Reaktion.

———. 2016. *Food and War in Mid-Twentieth-Century East Asia.* Routledge.

DAVIS, CHRISTINA L. 2003. *Food Fights Over Free Trade. How International Institutions Promote Agricultural Trade Liberalization.* Princeton: Princeton University Press.

DEES, BOWEN C. 1997. *The Allied Occupation and Japan's Economic Miracle: Building the Foundations of Japanese Science and Technology 1945–52.* London: Routledge.

DE MATOS, CHRISTINE. 2006. «Diplomacy Interrupted?: Macmahon Ball, Evatt and Labor's Policies in Occupied Japan». *Australian Journal of Politics & History* 52 (2): 188–201.

DESTLER, I. M. «United States Food Policy 1972–1976: Reconciling Domestic and International Objectives». In *International Organization* 32 (3): 617–653.

DOWER, JOHN W. 1979. *Empire and Aftermath: Yoshida Shigeru and the Japanese Experience, 1878–1954.* Cambridge, Massachusetts: Council on East Asian Studies, Harvard University.

———. 1999. *Embracing defeat: Japan in the wake of World War II.* New York: WW Norton & Company.

FARINA, FELICE. 2017. «Japan in the International Food Regimes: Understanding Japanese Food Self-Sufficiency Decline». In *Feeding Japan*, eds. Andreas Niehaus and Tine Walravens, 363–84. Cham: Springer International Publishing.

———. 2023. *La via della soia.* Sommacampagna: Cierre Edizioni.

FINN, RICHARD B. 1992. *Winners in Peace: MacArthur, Yoshida, and Postwar Japan.* Berkeley: University of California Press.

FOLLY, MARTIN. 2014. *Historical Dictionary of U.S. Diplomacy During the Cold War.* Lanham, Maryland: Rowman & Littlefield Publishers.

FORSBERG, AARON. 1995. «Eisenhower and Japanese Economic Recovery: The Politics of Integration with the Western Trading Bloc, 1952–1955». *The Journal of American-East Asian Relations* 5 (1): 57–75.

———. 1998. «The Politics of G.A.T.T. Expansion: Japanese Accession and the Domestic Political Content in Japan and the United States, 1948–1955». *Business and Economic History* 27 (1): 185–195.

FRANCKS, PENELOPE. 1998. «Agriculture and the state in industrial East Asia: the rise and fall of the food control system in Japan». *Japan Forum* 10 (1): 1–16.

———. 2006. *Rural Economic Development in Japan. From the Nineteenth Century to the Pacific War.* London: Routledge.

Frattolillo, Oliviero. 2019. *Reassessing Japan's Cold War. Ikeda Hayato's Foreign Politics and Proactivism During the 1960s*. London, Routledge.

———. 2014. *Il Giappone tra Est e Ovest. La ricerca di un ruolo internazionale nell'era bipolare*. Milano: FrancoAngeli.

Friedmann, Harriet. 1993. «The Political Economy of Food: A Global Crisis». *New Left Review*, I, no. 197: 29–57.

Friedmann, Harriet and Philip McMichael. 1989. «Agriculture and the State System: The Rise and Decline of National Agricultures, 1870 to the Present». In *Sociologia Ruralis* 29 (2): 93–117, 1989;

Fuchs, Steven J. 2007. «Feeding the Japanese: Food Policy, Land Reform, and Japan's Economic Recovery». In *Democracy in Occupied Japan: The U.S. Occupation and Japanese Politics and Society*, eds. Mark E. Di Caprio and Yoneyuki Sugita, 26–47. London: Routledge.

Fujihara, Tatsushi. 2018. *Kyūshoku no rekishi* [The history of school lunch]. Tōkyō: Iwanami Shoten.

George Mulgan, Aurelia. 1997. «The role of foreign pressure (gaiatsu) in Japan's agricultural trade liberalization». *The Pacific Review* 10 (2): 165–209.

———. 2006. *Japan's Agricultural Policy Regime.* London: Routledge.

Harding, Christopher. 2018. *Japan Story: In Search of a Nation, 1850 to the Present.* London: Penguin U.K.

Hirasawa, Akihiko. 2017. *Formation of Japan's food security policy: Relations with food situation and evolution of agricultural policies*, Papers and Reports, Norinchukin Research Institute Co., Ltd.

Hiratsuka, Masao. 2019. *Shashin de wakaru jiten no senryōshi: 1945nen 8gatsu—1952nen 5gatsu* [English title: History of occupied Japan: Aug.1945–May.1952]. Kyōto: Php Editors.

Ho, Alfred K. 1973. *Japan's Trade Liberalization in the 1960s*. New York: International Arts and Science Press.

Ho, Samuel Pao-San. 1984. «Colonialism and Development: Korea, Taiwan, and Kwantung». In *The Japanese Colonial Empire, 1895–1945*, ed. Raymond H. Myers and Mark R. Peattie, 347–399. Princeton: Princeton University Press.

Hopson, Nathan. 2020. «Ingrained habits: the "kitchen cars" and the transformation of postwar Japanese diet and identity». *Food, Culture & Society* 23 (5): 589–607.

Hosono Akio, Carlos Magno Campos da Rocha and Hongo Yutaka. 2016. *Development for Sustainable Agriculture. The Brazilian Cerrado*. New York: Palgrave MacMillan.

INOUE, MASAYUKI. 1988. «Daizu jiyūka 28 nen o kataru [Reflecting on 28 Years of Soybean Liberalization]». In *Aff: Agriculture, forestry, fisheries* 19 (12): 54–57.

ISHII, SUSUMU. 2004. «M.S.A. kyōtei to Nihon. Sengo kei Keizai shisutemu no keisei (2) [The M.S.A. Agreement and Japan: Formation of the Postwar Economic System]». *Gakushūin daigaku Keizai ronshū* 40 (4), 204: 295–313.

ISHIKAWA, HIROTOMO. 1981. *Kokumotsu mejā: Shokuryō senryaku no "kage no shihaisha"* [Grain Majors: Rulers of Food Strategy]. Tokyo: Iwanami Shoten.

ITAMI, HIROYUKI. 2019. «Chapter 4. Economic Growth and Japanese Management». In *Seven Chapters on Japanese Modernization*. J.I.C.A.-Open University of Japan, BS231.

ITŌ, ATSUSHI. 2021. «Japanese Market Development Programs under PL 480 in the late 1950s: Focus on Wheat and Rice». *Agricultural History* 95(2): 245–275.

———. 2020. «PL480 taitoru II o meguro nichibei kōshō—Gakudō-fuku keikaku no dannen to gakkō kyūshoku muke zōyo no seiritsu [The Negotiations between the U.S. and Japan concerning PL 480 Title II: The Abandonment of a School Children's Clothing Program and the Formation of a Grant for a School Lunch Program]». In *Nōgyō keizai kenkyū* 92 (2): 165–177.

———. 2021. «Dai 1 ji PL480 taitoru I kyōtei o meguru nichibei kōshō, 1954–55 [Negotiations between Japan and the United States regarding the First PL480 Title I Agreement, 1954–55]». In *Rekishi gaku kenkyū* 3: 1–15.

———. 2022. «Dai 2ji PL480 kyōtei taitoru I kyōtei o meguru Nichibei kōshō (1955–56). Nichibei nikokukan kankei o koete [Negotiations between Japan and the United States regarding the Second PL480 Title I Agreement, 1955–56: beyond bilateral relations]». *Shakai Keizaishi-gaku*, 88 (1): 33–55.

———. 2022. «M.S.A. yojō nōsanbutsu kyōtei o meguru nichibei negotiations 1953–60—Dai 402 jō kyōtei kōshō ni chakumoku shite [The Negotiations between the U.S. and Japan Concerning Surplus Agricultural Commodities Agreements under the Mutual Security Act, 1953–60: A Focus on Section 402 Negotiations]». In *Nōgyō keizai kenkyū* 94 (3): 208–213.

———. 2023. «Dai 3ji dai 4ji PL480 kyōtei taitoru I kyōtei o meguru Nichibei kōshō (1956–57 nen). Seiritsu shinakatta kyōtei no kōshō kara no shōsha [Negotiations between Japan and the United States regarding the Third and Fourth PL480 Title I Agreement, 1956–57: from the viewpoint of rejected agreements]». *Shakai Keizaishi-gaku* 89 (1): 27–52.

JOHNSTON, BRUCE F. 1953. *Japan Food Management in World War II.* Wisconsin: Stanford University Press.

KAWAI, KAZUSHIGE. 2000. *Osorubeki "yunyū kome" senryaku: WTO kyōtei kara kome to tanbo wo mamoru tame ni* [The Terrifying "Imported Rice" Strategy: Protecting Rice and Rice Fields from WTO agreement]. Tokyo: Gōdō.

Kishi, Yasuhiko. 1996. *Shoku to nō no sengoshi* [The History of Food and Agriculture in the Postwar]. Tōkyō: Nihon keizai shinbunsha.

Kusakari Hitoshi. 2009. «Sengo no shokuryō shōhi to kakei no yakuwari [Postwar Food Consumption and the Role of Household Budgets]». *Japanese journal of research on household economics* (83): 12–25.

Kushnar Barak. 2012. *Slurp! A Social and Culinary History of Ramen—Japan's Favorite Noodle Soup.* Leiden: Global Oriental.

Lanna, Noemi, and Marisa Siddivò, eds. 2023. *The Challenge Ahead: The Politics of Food Security in China, India Japan and South Korea*. Naples: UniorPress.

Li Enmin. 1999. «Nihon no keizai anzen hoshō wo motomete. 1970 nendai ni okeru keidanren no shigen gaikō» [In Pursuit of Japan's Economic Security: Keidanren's Resource Diplomacy toward China during the 1970s]. In *Kenkyū Ronshū* 8: 67–116.

McDonald, Bryan L. 2017. *Food Power: The Rise and Fall of the Postwar American Food System*. New York: Oxford University Press.

McMichael, Philip. 2009. «A Food Regime Genealogy». In *The Journal of Peasant Studies* 36 (1): 139–169.

Meyerson, Christopher C. 2003. *Domestic Politics and International Relations in U.S.-Japan Trade Policymaking. The G.A.T.T. Uruguay Round Agriculture Negotiations*. London: Palgrave McMillan.

Miyazawa Kiichi. 1997. *Sengo seiji no shōgen* [Testimony of postwar politics]. Tokyo: Yomiuri shinbunsha.

Moen, Darrell G. 1999. «The Postwar Japanese Agricultural Debacle». In *Hitotsubashi Journal of Social Studies* 31 (1): 29–52.

Moore, Joe. 1985. «Production control: Workers' control in early postwar Japan». *Bulletin of Concerned Asian Scholars* 17 (4): 2–26.

Nakamura, Hiroshi. 1961. «The Japanese Soybean Market». In *Illinois Agricultural Economics*, 1 (2): 7–13.

Nakamura, Teiji. 2021. *Japan Nutrition*. Singapore: Springer Nature.

Niehaus, Andreas, and Tine Walravens, eds. 2017. *Feeding Japan: The Cultural and Political Issues of Dependency and Risk*. Palgrave Macmillan.

Ōmameuda Minoru. 1993. *Kindai nihon no shokuryō seisaku. Taigai izon beikoku kyōkyū kōzō no hen'yō* [Food policies in modern Japan: Transformation of the structure reliant on imported rice supply]. Kyoto: Minerva Shobō.

Paarlberg, Robert L. 1980. «Lessons of the grain embargo». In *Foreign Affairs* 59 (1): 144–162.

PORTER, ROGER B. 1984. *The U.S.-U.S.S.R. Grain Agreement.* Cambridge: Cambridge University Press.

REISCHAUER, EDWIN O. AND JANSEN, MARIUS B. 1999 (original printing 1977). *The Japanese Today. Change and Continuity.* Cambridge: Harvard University Press.

RILEY, BARRY. 2017. *The Political History of American Food Aid: An Uneasy Benevolence.* New York: Oxford University Press.

RIX, ALAN. 1979. «The Mitsugoro Project: Japanese Aid Policy and Indonesia». In *Pacific Affairs* 52 (1): 42–63.

———. 2011. *Japan's Economic Aid. Policymaking and Politics.* London: Routledge.

SAITŌ, MINAKO. 2015. *Senka no reshipi. Taiheiyō sensō-ka no shoku wo shiru* [Recipes from the War: Understanding Food During the Pacific War]. Tokyo: Iwanami Shoten.

SCALAPINO, ROBERT A. 1967. *The Japanese Communist Movement, 1920–1966.* Santa Monica: University of California Press.

SHIMAMOTO, MAYAKO, ITŌ, KOJI, AND SUGITA, YONEYUKI. 2015. *Historical dictionary of Japanese foreign policy.* Lahnam, Maryland: Rowman & Littlefield.

SHIMIZU, KAZUHIKO. 2015. «"Mohaya 'sengo' de wa nai" to iu shakaiteki kioku no kōsei katei» [The compositional process of the social memory that "it is no longer the 'postwar'"]. In *Edogawa Daigaku kiyō* [Edogawa University bulletin] 25 (March): 195–206.

«Shokuryō mēdē to Tennō purakādo jiken (2)—Matsushima Matsutarō-shi ni kiku [The Food May Day and the Placard Incident (2)—Interview with Matsushima Matsutarō]». 2003. *Ohara Shakai Mondai Kenkyū-sho Zasshi* [Journal of the Ōhara Institute for Social Research], n. 535 (June): 55–71.

SOLT, GEORGE. 2014. *The Untold History of Ramen: How Political Crisis in Japan Spawned a Global Food Craze,* Berkley: University of California Press.

SUZUKI, NOBUHIRO. 2013. *Shoku no sensō. Beikoku no wana ni ochita Nihon* [The War of Food: Japan Caught in the American Trap]. Tokyo: Bunshun.

SUZUKI, TAKEO. 2003. *Amerika no komugi senryaku to nihonjin no shokuseikatsu* [The American Wheat Strategy and Japanese Food Habits]. Tokyo: Fujiwara Shoten.

TAIHEIYŌ SENSŌ KENKYŪ-KAI. 2007. *G.H.Q. no mita Nippon* [Japan seen by the G.H.Q.]. Sekaibunkasha.

TAKAGI, SHINJI. 1995. *From recipient to donor: Japan's official aid flows, 1945 to 1990 and beyond,* Essay in International Finance. Princeton: Princeton University.

———. 1997. «Japan's Restrictive System of Trade and Payments Operation, Effectiveness, and Liberalization, 1950–1964». I.M.F. Working Paper.

———. 2015. *Conquering the Fear of Freedom: Japanese Exchange Rate Policy since 1945*. Oxford: Oxford University Press.

Takashi, Kazuhiro. 2012. «Ikeda seiken-ki ni okeru bōeki jiyūka to nashonarizumu [Trade Liberalization and Nationalism during the Ikeda Administration]». *Kokusai seiji* 170: 46–60.

Takita, Kenji. 2015. "Kokusai shakai to Amerika no senryō tainichi keizai enjo—G.A.R.I.O.A., E.R.O.A. enjo wo chūshin toshite [The International Community and Postwar U.S. Economic Assistance to Japan—A Focus on G.A.R.I.O.A. and E.R.O.A.]". *Hōgaku shinpō,* March 2015: 326–345.

Tatara, Toshio. 1999. *Kyūen no busshi wa taiheiyō o koete: Sengo Nihon to Lala no katsudō* [Relief Supplies Cross the Pacific: Postwar Japan and L.A.R.A.'s Activities]. Tōkyō: Hoken fukushi kōhō kyōkai.

Tsurumi, Shunsuke. 2011. *A Cultural History of Postwar Japan 1945–1980*. London & New York: Routledge.

Unno, Hiroshi. 2016. *Shokuryō mo daijōbu nari. Kaisen to shūsen no ketsudan to shokuryō* [English title: Starting and Ending the War (1941–1945). From the Perspective of Food Logistics]. Tōkyō: Nōrin tōkei shuppan.

Welfield John. 2013. *An Empire in Eclipse: Japan in the Postwar American Alliance System: A Study in the Interaction of Domestic Politics and Foreign Policy*. London: Bloomsbury Publishing.

Yamata, Samuel Hideo. 2013. «The "Food Problem" of Evacuated Children in Wartime Japan, 1944–1945». In *Food and War in Mid-Twentieth Century East Asia*, eds. Katarzyna J. Cwiertka, 131–48. Farnham: Ashgate.

www.ingramcontent.com/pod-product-compliance
Ingram Content Group UK Ltd.
Pitfield, Milton Keynes, MK11 3LW, UK
UKHW020842290926
15470UKWH00004B/1505